Self-propelled anti-tank guns had first been employed by the German army during the 1940 French campaign when four independent battalions were equipped with the Panzerjäger I, a converted Pzkpfw I ausf B mounting a 4.7cm Pak 38 gun, a captured Czech weapon. These proved so successful that a new formation was raised for service in North Africa and six battalions took part in Operation Barbarossa, the invasion of the Soviet Union. By the summer of 1942 most had either been destroyed or phased out of service but the basic concept was thought to be sound and development continued with more powerful guns and heavier chassis (1).

But these vehicles were essentially a stop-gap measure and although details of the designs varied, they could all be described as a standard anti-tank gun mounted on an existing chassis which was modified by the addition of raised armoured sides. They were, in fact, a quick and cheap solution to the problem of the enemy's growing superiority in not only numbers but also weapons technology. In their intended role as a mobile platform for an anti-tank weapon they were devastatingly effective but when forced to manoeuvre, as they often were, their high silhouette and open top and rear proved to be a decided disadvantage.

Another self-propelled design, the Sturmgeschütz III, had originally been designed as an infantry support weapon and although this role was never totally abandoned, assault guns were increasingly employed as tank killers. In early 1942 the introduction of the 7.5cm L/43 gun (2), which was soon replaced by a still more powerful weapon, meant that the Sturmgeschütz was capable of taking on and destroying the strongest enemy tanks. Sturmgeschütz battalions, whether independent or permanently attached to divisions, were part of the artillery inspectorate and General Guderian, as Generalinspekteur der Panzertruppen, regularly complained to Hitler that these vehicles essentially fulfilled the role of tanks and should therefore come under his control (3). Hitler prevaricated and, in a compromise solution which was typical of the Führer, authorised the development of a dedicated tank hunter, or Jagdpanzer.

In September 1942 the Heereswaffenamt promulgated a specification for what was considered at the time, a development of the Sturmgeschütz (4).

The new design was to be armed with the 7.5cm L/70 gun and protected by 100mm thick frontal armour while the sides were of 40 and 50mm thickness. It was to incorporate the lowest possible silhouette and be capable of a top speed of 25 kilometres per hour. The firm of Altmärkische Kettenwerk GmbH (Alkett) of Berlin was the first to submit a design which was based on the chassis of a Pzkpfw IV and armed with either a 7.5cm L/70 or 10.5cm gun. For some unknown reason the Alkett proposal was never accepted, although Hitler is said to have been satisfied with the design, and it was stipulated that any future development was to be based on components of the Pzkpfw III tank. But as any available hulls and chassis were earmarked for the Sturmgeschütz III programme this idea proved impractical.

A proposal from Krupp AG, referred to as the Panzerjäger IVb E39, was eventually dismissed as it would have required the employment of a new suspension system which was only then in the early stages of development and it was felt that this would have disrupted production of essential weapons systems. At the same time a

Notes

1. The fundamentals of the design could be said to have culminated in the Sdkfz 164 Nashorn, based on the Pzkpfw IV chassis and armed with the 8.8cm Pak 43, which remained in service until the end of the war.

2. The letter L refers to the calibre of a particular gun and the number is arrived at by dividing the length of the barrel by its internal diameter. For example, 360cm divided by 7.5cm equals 48. Generally speaking, the longer the barrel the greater the effective range.

3. A program to provide a company of assault guns to the Panzerjäger battalion of each infantry division was instituted from mid-1943 and although many of these units were equipped with Sturmgeschütz III and Sturmgeschütz IV assault guns they were still considered part of the Panzertruppen.

4. The Heereswaffenamt was the department of the army that was concerned with the development and manufacture of weapons and vehicles. The Heereszeugamt controlled the supply of weapons, ammunition and other stores.

Said to have been photographed during the first weeks of October 1944, this early production Jagdpanzer IV L/48 tank destroyer may be one of the vehicles which SS-Panzerjäger-Abteilung 17 lost in this area of north-eastern France in the previous month. Note that the rack on the rear hull for the length of spare track does not extend to the edges of the vehicles, an early production feature.

The Vogtländische Maschinenfabrik AG (Vomag) factory at Plauen photographed in the spring of 1945. The town was heavily damaged during the last year of the war with fourteen raids occurring within a four-week period from mid-September to early October 1944. It was here that production of the Jagdpanzer IV L/48 and Panzer IV L/70(V) took place and the plant was not seriously damaged until March 1945, continuing vehicle assembly until that time. The factory had its own extensive system of air raid shelters, emergency electricity and water supplies and employed 8,000 people at the height of production in late 1944. What remained of the complex was destroyed by the Soviets in 1946.

design submitted by Vogtländische Maschinenfabrik AG (Vomag) utilised the almost unaltered chassis of the Pzkpfw IV ausf F tank and Hitler, after viewing a wooden mock-up, gave his approval for work to continue on the prototype. By late 1943 a mild steel version had been completed and presented to the Führer, who agreed that production should procede. The initial design included rounded front edges and firing ports on each side of the fighting compartment but both features were dropped from the production version in favour of a much simpler design. In addition Hitler's demand that the 7.5cm L/70 gun be fitted to the new tank destroyers could not be realised as these were all needed in the production of the Pzkpfw V Panther and the L/48 gun, the same weapon used in the Sturmgeschütz III programme, was employed. The maximum armour thickness was reduced from 100mm to 60mm but the angle of the plates was increased which compensated somewhat.

The first production vehicles were delivered from Vomag as Sturmgeschütz neue Art, or new model assault gun, and in March 1944 the first Jagdpanzers were allocated to operational units and training establishments.

Hitler's interest in the L/70 gun had not been diminished and by early April 1944 a Jagdpanzer IV was fitted with the more powerful gun as a trial and subsequently ordered into production. The Vomag firm, which was already building the Jagdpanzer IV, was contracted to construct the new design which was eventually designated Panzer IV L/70(V). Shortfalls in production at Vomag saw a similar design from Alkett introduced in August 1944 as the Panzer IV L/70(A). The development of these vehicles and the technical aspects peculiar to each are discussed in greater detail from page 58.

Readers may be confused by the use of the terms Panzerjäger and Jagdpanzer in connection with the vehicles examined in our study. Both terms seem to have been used at various times in official wartime documents to describe these tank destroyers and I have endeavoured to at least achieve some consistency here. When referring to the series in general I have used the term Jagdpanzer but when specifically describing the initial design I have employed the full title of Jagdpanzer IV L/48 or Jagdpanzer IV ausf F.

Another cause of some confusion are the numerous modern inventions such as Zwischenlösung. This term, which is often applied to the Panzer IV L/70(A) and means interim design, was never used during the war.

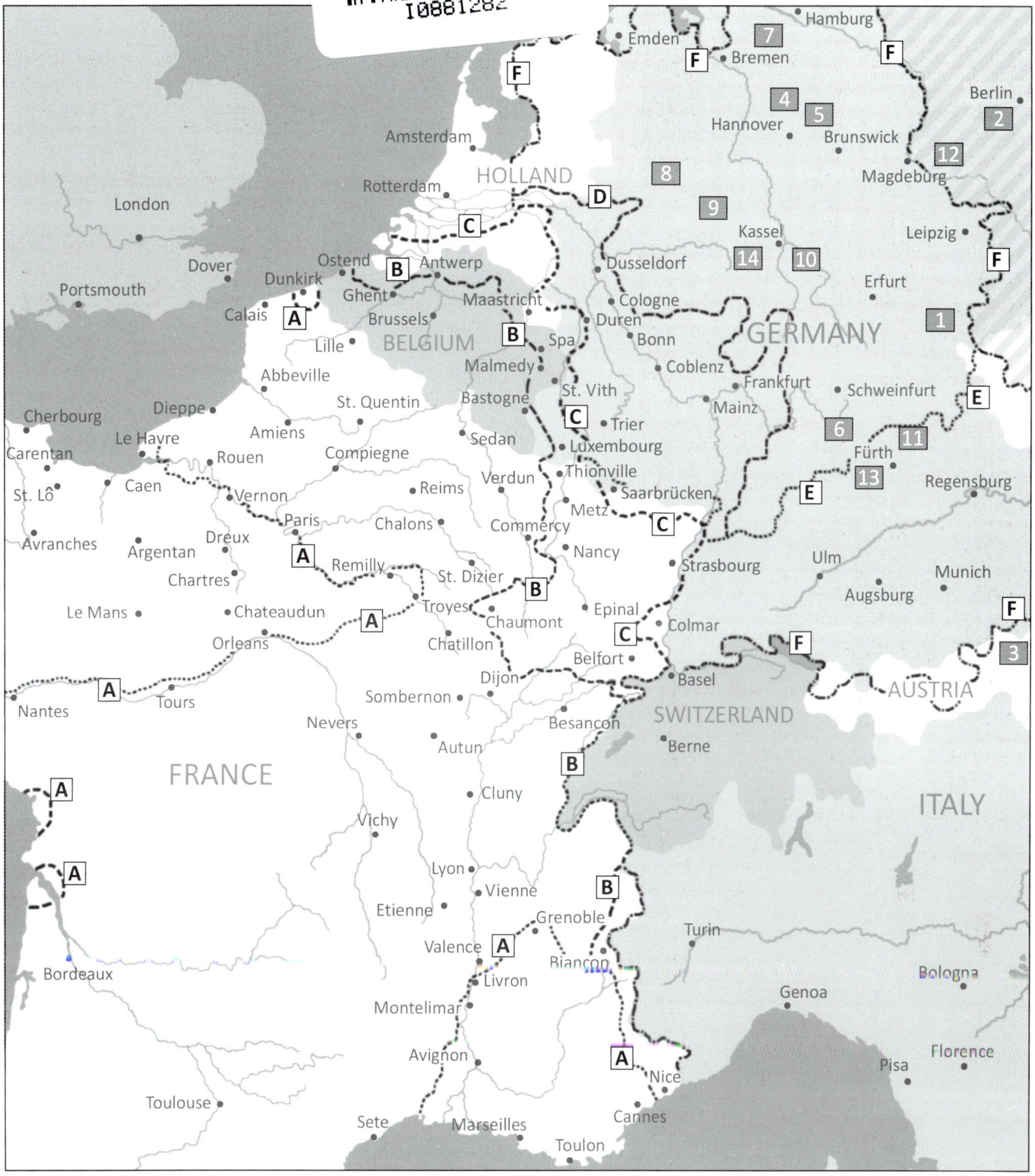

With the Allied breakout from Normandy in July and August 1944 the front moved rapidly eastwards and by mid-September ran from just north of Ostend along the Dutch-Belgian border to Maastricht, then south to the outskirts of Aachen. From there the line ran south through Luxembourg, following the Moselle through Thionville, then continuing south to the east of Epinal and Belfort before turning sharply towards the Swiss frontier. By 16 December 1944, and the commencement of Wacht am Rhein, the Americans had overrun most of Alsace and Lorraine and the British and Canadians had advanced as far as Nijmegen in the north. The latter was, however, the ground gained during Operation Market-Garden in late September and the centre of the front remained basically unchanged. The national borders depicted here are those of 1945 and although I have endeavoured to use the correct contemporary place names throughout the text, some anglicised versions are shown here. The approximate location of the front line is shown as follows: A. 26 August 1944. B. 25 September 1944. C. 5 January 1945. D. 24 March 1945. E. 18 April 1945. F. 7 May 1945. Note that the front as shown for 7 May 1945 is in fact the line of demarcation as agreed with the Soviets and some US Army units had penetrated beyond this, particularly in southern Germany and Czechoslovakia, only to be ordered to withdraw. The shaded area to the east of the 7 May line was occupied by the Red Army at that time although the area to the south was still in contention and the last German units here did not surrender until 11 May 1945.

Places of importance mentioned in the main text are indicated by the following numbers: 1. Vogtländische Maschinenfabrik AG (Vomag), Plauen. 2. Altmärkische Kettenwerk GmbH (Alkett), Berlin. 3. Nibelungenwerke, St. Valentin. 4. Panzertruppen-Schule Bergen. 5. Schiess-Schule der Panzertruppen Putlos. 6. Panzerlehrgang Erlangen. 7. Lehrtruppe der Panzertruppen-Schule Fallingbostel. 8. Truppenübungsplatz Münster. 9. Truppenübungsplatz Sennelager. 10. Truppenübungsplatz Ordruf. 11. Truppenübungsplatz Grafenwöhr. 12. Heeres-Zeugamt Altengrabow. 13. Heeres-Zeugamt Grafenwöhr. 14. Heeres-Zeugamt Sennelager.

This battered Panzer IV/70(A) of Panzer-Regiment 2 was captured in running order, although damaged, by French Army units near Rudlin, west of Colmar, in December 1944. This vehicle is also shown and discussed on page 25 of the Camouflage & Markings section of this book.

Notes

1. Even this classification is not strictly accurate and as many as eight different titles are known to have been used in official documents.

During 1944 Sturmgeschütz assault guns and Jagdpanzer tank destroyers were increasingly used to replace losses within Panzer regiments and battalions and by the end of the year mixed, or Gemischte, battalions were common. Listed below are the Panzer units that operated on the Western Front that were allocated tank destroyers based on the Pzkpfw IV chassis. Many records from this period are incomplete or sometimes contradictory and, as readers of this series will know, allocated does not always mean delivered. In the instances where the available information is imprecise or sources disagree I have tried to make this clear.

As a matter of convenience I have used the term Jagdpanzer IV in place of the more correct, but longer, Jagdpanzer IV 7.5cm Pak 39 L/48 (1). I have used the word Jagdpanzer, without any identifying suffix, as a generic term to describe any of the three models mentioned here.

Panzer-Regiment 2. In mid-November 1944 the regiment's II.Abteilung received eleven Panzer IV/70(A) and all were allocated to 7.Kompanie under the command of Oberleutnant Walther Oehme. The company was also equipped with six Pzkpfw IV ausf J tanks and formed into an ad-hoc Kampfgruppe with elements of 1.Kompanie, schwere Panzerjäger-Abteilung 520 and 1.Kompanie, Schwere Panzerjäger-Abteilung 93.

Fighting in the Colmar Pocket, 7.Kompanie reported that three of the tank destroyers had been lost by 16 January 1945 but claimed the destruction of fifty enemy armoured vehicles. On 15 February Oberleutnant Oehme and his men were ordered to Grafenwöhr in Germany to rest and refit and the tank destroyers were taken over by 2.Kompanie, schwere Panzerjäger-Abteilung 93 which was attached to Panzer-Brigade 106 (see below).

Panzer-Regiment 33. Attached to 9.Panzer-Division, the regiment was rebuilt, largely in the field, after the disastrous losses suffered in the final stages of the Normandy battles. By the autumn of 1944 the division had been reduced to a Kampfgruppe and fought in the battles for Geilenkirchen and Aachen, and later took part in the Ardennes Offensive and the fighting in the Eifel region, most notably the defence of the Erft River crossings.

In March 1945, the surviving elements of 9.Panzer-Division fought in the battles for the Remagen bridgehead and on 14 April 1945 II.Abteilung, Panzer-Regiment 33 received ten Panzer IV/70(V) tank destroyers. The division was trapped in the Ruhr Pocket and although some elements managed to escape it was officially disbanded on 26 April 1945. See also Panzerjäger-Abteilung 50 below.

Panzer-Regiment Grossdeutschland. See Führer-Begleit-Brigade below.

Panzer-Abteilung 5. Attached to 25.Panzergrenadier-Division, the battalion, with most of the division, was almost completely destroyed in the fighting in Russia during the summer of 1944 with the surviving elements used to form Panzer-Brigade 107.

From October 1944 the division was rebuilt and Panzer-Abteilung 5 was issued eleven Panzer IV/70(V) tank destroyers

...........text continued on page 7

PANZER-REGIMENT, NOVEMBER 1944

All formations of the German Army were organised according to detailed tables of establishment referred to as Kriegstärkenachweisung (KstN) which showed the official composition of a unit, listing the exact number of personnel and type of equipment from vehicles to small arms. On 3 August 1944 Oberkommando des Heeres, or OKH, issued orders that all Army Panzer divisions were to be reorganised according to a standardised configuration referred to as Gliederung Panzer-Division 44. This was to be accomplished using the instructions which had been issued earlier on 1 April 1944 with the stipulation that any changes were to be completed by divisions within "...the bounds of their available personnel and material." In November 1944 further instructions were issued to be used for new formations or units undergoing rebuilding and these allowed for companies to be equipped with either the Panther, Pzkpfw IV tank or Panzer IV/70 (A) tank destroyers. The companies could be made up of seventeen, fourteen or ten vehicles each, as shown in the diagram below. It is not known how many formations, if any, were fully formed using the November instructions and it should be remebered that these orders were promulgated at a time when production of the Panzer IV/70 (A) was expected to replace that of the Pzkpfw IV tank.

Stab und Stabskompanie eines Panzer-Regiments KstN 1103 von 1.11.1944

Regiments Stab

Stabskompanie Gruppe Führer — 3 x Panzerbefehlswagen V Panther

Aufklärungs Zug — 5 x Pzkpfw IV

Stab und Stabskompanie (frei Gliederung) einer Panzer-Abteilung IV KstN 1107 (fG) ausf A von 1.4.1944

Gruppe Führer — 3 x Panzerbefehlswagen IV or Pzkpfw IV

Aufklärungs Zug — 5 x Pzkpfw IV

Stab und Stabskompanie (frei Gliederung) einer Panzer-Abteilung Panther KstN 1107 (fG) ausf B von 1.4.1944

Gruppe Führer — 3 x Panzerbefehlswagen V or Pzkpfw V Panther

Aufklärungs Zug — 5 x Pzkpfw V Panther

Both KstN for the battalion headquarters shown above were amended to Stab und Stabskompanie (frei) einer Panzer-Abteilung IV ausf A, Panzer-Abteilung Panther ausf B, Panzer-Sturmgeschütz-Abteilung und Sturmpanzer-Abteilung ausf C and schwere Panzer-Abteilung Tiger ausf D, all included in KstN 1107(fG) ausf von 1.11.1944.

Panzerkompanie | **Panzerkompanie** | **Panzerkompanie**

Depending on the availability of suitable vehicles, companies could be formed with seventeen, fourteen or eleven tanks or tank destroyers and the KstN issued for November 1944 are shown below.

Panzerkompanie Panther oder Panzer IV (17 Pzkpfw) (frei Gliederung) KstN 1177 (fG) ausf A von 1.11.1944

Kompanie Gruppe Führer — 2 x Pzkpfw V Panther OR 2 x Pzkpfw IV OR 2 x Panzer IV/70(A)

1.Zug — 5 x Pzkpfw V Panther OR 5 x Pzkpfw IV OR 5 x Panzer IV/70(A)

2.Zug — As for 1.Zug

3.Zug — As for 1.Zug

Panzerkompanie Panther oder Panzer IV (17 Pzkpfw) (frei Gliederung) KstN 1177 (fG) ausf B von 1.11.1944

Kompanie Gruppe Führer — 2 x Pzkpfw V Panther OR 2 x Pzkpfw IV OR 2 x Panzer IV/70(A)

1.Zug — 4 x Pzkpfw V Panther OR 4 x Pzkpfw IV OR 4 x Panzer IV/70(A)

2.Zug — As for 1.Zug

3.Zug — As for 1.Zug

Panzerkompanie Panther oder Panzer IV (17 Pzkpfw) (frei Gliederung) KstN 1177 (fG) von 1.11.1944

Kompanie Gruppe Führer — 2 x Pzkpfw V Panther OR 2 x Pzkpfw IV OR 2 x Panzer IV/70(A)

1.Zug — 3 x Pzkpfw V Panther OR 3 x Pzkpfw IV OR 3 x Panzer IV/70(A)

2.Zug — As for 1.Zug

3.Zug — As for 1.Zug

Both photographs on this page show the same Panzer IV/70(A) in the streets of Mittlewihr, just north of Colmar. This vehicle was disabled when 7.Kompanie, Panzer-Regiment 2, which was at that time attached to Panzer-Brigade 106 Feldherrnhalle, attempted to take the town in early December 1944. The wire mesh Schürzen or Drahtgeflect and its supporting rail is missing from the driver's side of the vehicle although the skirt for the other side is still in place.

In the original print of both images it is possible to discern a solid black Balkenkreuz at the very top of the crew compartment, to the right of the gun mantlet while the marking on the side has the standard white outline. Note the armoured cover for the hull machine gun in the open position. The ferocity of the fighting here, which lasted until January 1945, is evident in the destruction shown in these photographs and the village was never rebuilt, with only a small part of the medieval church and a wall remaining today.

...........*continued from page 4*

in place of its authorised allocation of fifteen Pzkpfw IV tanks. These were all assigned to 4.Kompanie and as the division received very few new vehicles all eleven tank destroyers probably came from the disbanded Panzer-Brigade 107. This would at least explain why just two were reported as fully operational in December 1944 and only five took part in Operation Nordwind in January 1945.

In February the division was transferred to the Eastern Front. The suggestion in some accounts that the division's Panzerjäger-Abteilung 25 was equipped with either Jagdpanzer IV or Panzer IV/70 (A) tank destroyers is almost certainly incorrect.

Führer-Grenadier-Brigade. Raised in June 1944 from elements of the Wolfschanze security detachment and replacements for the Grossdeutschland division, elements of the brigade fought on the Eastern Front until September when it was transferred to Fallingbostel in Germany.

The 5.(Panzer) Kompanie was reorganised as a full battalion and referred to as III.Panzer-Abteilung, Führer-Grenadier Brigade. In September the battalion received eleven Panzer IV/70(V) tank destroyers and these were all allocated to 4.Kompanie (1). The brigade fought in East Prussia until November 1944 when it was transferred to the West and subordinated to 5.Panzerarmee. Commanded by Oberst Hans-Joachim Kahler, the Führer-Grenadier-Brigade took part in the Ardennes Offensive, most notably at Vianden and around Martelange, but was committed far too late to have any impact on the battle. The brigade was withdrawn from the front in mid-January 1945 and upgraded to divisional status at the end of the month with III.Abteilung reorganised as Panzer-Regiment 101. The division was sent to the Eastern Front in February and took part in the recapture of Lauban.

Führer-Begleit-Brigade. Formed at the same time as the Führer-Grenadier-Brigade, this unit was initially supposed to contain a single Panzer company equipped with seventeen Pzkpfw IV ausf J tanks and five Panzer IV/70(A) tank destroyers shipped from the Heereszeugamt in September and October 1944. Prior to this, in August 1944, II.Abteilung, Panzer-Regiment Grossdeutschland had been withdrawn from the fighting in the East to be reorganised but the forty-five tanks that had been promised were never delivered and in their place the battalion was equipped with seven Pzkpfw IV ausf J tanks and thirty-eight Panzer IV/70(A) tank destroyers, all allocated in November.

By the beginning of December 1944, the battalion had been attached to the newly formed Führer-Begleit-Brigade and operated as I.Abteilung of the brigade's Panzer regiment. Commanded by Oberst Otto Remer, the brigade was subordinated to XLVII.Panzerkorps during the Ardennes Offensive and from 18 December 1944 took part in the fighting around St Vith and later at Bastogne.

On 26 January 1945 the brigade was reorganised as the Führer-Begleit-Division although it never reached its full authorised strength. The division was sent to the East and remained there until the end of the war, although some survivors, including Remer, were able to surrender to units of the US Army.

Notes

1. In some accounts these are erroneously referred to as Panzer IV/70(A). All eleven were on hand in December. Confusingly the allocation records often include the Jagdpanzer IV and Panzer IV/70(A) with the Pzkpfw IV lists as 'Pz.IV.lg' and the Panzer IV/70(A) shipments are sometimes included with the Panzer IV/70(V). I am indebted to the many researchers that have managed to establish the correct figures.

Abandoned in an Esso filling station, this Panzer IV/70(A) tank destroyer was assembled at some time after September 1944 when the Drahtgeflecht wire mesh screens began to replace the hull Schürzen in production. Note the Flammentöter exhaust mufflers, with which the great majority of these vehicles were fitted, and the armoured cover for the FuG 8 command radio on the rear of the superstructure. Note also the Kugellafette Vorsatz P mount on the loader's hatch which was authorised in November 1944 but fitted to very few vehicles. The building in the background, fully visible in the original print, is still standing today and identifies the location as the Rue de Verdun in Wittenheim, just north of Mulhouse. The town was captured by the Allies on 22 November 1944 and if the location is correct, and I believe it is, this vehicles may be one of the eleven tank destroyers allocated to Panzer-Regiment 2 earlier that month.

This early production Panzer IV/70(V) tank destroyer is fitted with the cylindrical exhaust muffler, four return rollers and is coated in Zimmerit indicating that it was assembled in August or early September 1944. This vehicle is almost certainly one of the Panzer-Brigade 107 vehicles lost in the battles of autumn 1944 and was tested extensively by the Allies.

The history and organisation of the independent Panzer brigades formed from July 1944 is covered in some detail in *TankCraft 24 Panther Tanks: German Army Panzer Brigades, Western and Eastern Front, 1944-1945* but limitations of space prevent us from repeating all the information here. Therefore I have concentrated on the events affecting the Jagdpanzers that were issued to the formations which served in the West.

Panzer-Abteilung 2105. Attached to Panzer-Brigade 105, the battalion was established in July 1944 at Bamberg under the auspices of Panzer-Ersatz-Abteilung 35 and the tank companies were trained on the Pzkpfw V Panther at Bamberg and Erlangen. In early September 1944 the brigade was transferred by rail to the area between Brussels and Namur in Belgium and was immediately caught up in the general retreat and the tanks of Panzer-Brigade 105 went into combat for the first time on 7 September 1944 in an attack made east of Leuven, close to Brussels. The eleven Panzer IV/70(V) tank destroyers, which were to equip the fourth company of the Panzer battalion, were not despatched until 4 September 1944 and by the time the company was able to join its parent formation the brigade had been at the front for some time. The brigade took part in the battles in eastern Belgium, most notably in the defence of Fleron and in a counterattack at Malmedy, and from mid-September supported 9.Panzer-Division in the fighting on the Westwall defensive line south of Aachen, losing many of its armoured vehicles and over half the Panzergrenadier battalion. On 23 September 1944 the brigade was disbanded and most of the survivors were integrated into the units of 9.Panzer-Division.

Panzer-Abteilung 2106. Attached to Panzer-Brigade 106, the battalion was formed on 24 July 1944. Most of the brigade staff and the Panzergrenadier battalion were taken from Panzer-Regiment 11 and the Ersatzheer, or replacement army, but much of the combat elements were built around the remnants of Panzergrenadier-Division Feldherrnhalle and Panzer-Ersatz- und Ausbildungs-Abteilung Feldherrnhalle and the brigade also received the title (1).

Panzer-Abteilung 2106 contained three tank companies equipped with Pzkpfw V Panther tanks and a fourth company which was allocated eleven Panzer IV/70 (V) tank destroyers shipped from the Heereszeugamt on 29 August 1944. The battalion's 4.Kompanie could have spent little or no time in training as the last of their Panzer IV/70(V) may have arrived after the brigade went into action for the first time in the defensive operations at Briey, south-west of Thionville, and the withdrawal towards Trier. Although the

Notes

1. The honour title Feldherrnhalle was originally bestowed on Infanterie-Regiment 271 in 1942 in recognition of the large number of former Sturm-Abteilung (SA) men within its ranks. The title refers to a monument commissioned in 1841 to honour the Bavarian army but later appropriated by the Nazis as the site where fourteen of Hitler's entourage were killed by the police during the failed Beer Hall Putsch of 1923. During the war a number of army units received the title, including 13.Panzer-Division and schwere Panzer-Abteilung 503, but by 1944 the connection with the SA had largely been lost.

brigade's initial operations were largely successful, with the Panzer battalion claiming the destruction of twenty-six American tanks and armoured cars in a single day near Oberkorn on the French-Luxembourg border, they were extremely costly and in its first day of combat twenty-one tanks and tank destroyers and over sixty halftracks were lost.

But within days the brigade supported a series of counterattacks made by 3.Panzergrenadier-Division against the Moselle bridgehead south of Metz and remained in this area until the end of September when it was moved to the north-west of Colmar, and then to Belfort. In late October Panzer-Brigade 106 supported 21.Panzer-Division and elements of 16.Volksgrenadier-Division in a counterattack near Raon where the Americans had broken through the German defensive line near Mortagne. During November 1944 the brigade was involved in defensive actions as part of the operational reserve of 1.Armee along the so-called Falkenberg Line, south-east of Metz, and also at Bechingen and Saint-Avold.

On 8 December the brigade reported that four Jagdpanzers were on hand together with just ten Panthers and after the battles around Kayserberg the brigade was withdrawn to Soultz-Haut-Rhin, north of Mulhouse, and held in reserve but had returned to the front by the middle of January to take part in the counterattacks on the Alsace bridgehead. On 17 February 1945 the brigade staff received orders to take over eight Panzer IV/70(A) tank destroyers which had been left near the city of Colmar with 2.Kompanie, schwere Panzerjäger-Abteilung 93 (see Panzer-Regiment 2 above) and proceed to Cologne. The 15 March 1945 report prepared for the Inspekteur der Panzertruppen states that seven of these vehicles were still on hand, of which four were fully operational.

From 6 April 1945 the surviving elements of the brigade were incorporated into the newly raised Panzer-Division Clausewitz. Initially the Panthers and Panzer IV/70(A) tank destroyers were organised into separate companies of Gemischtes-Panzer-Regiment Clausewitz but on 17 April they were amalgamated into a single mixed company.

On 22 April 1945, after taking part in the heavy fighting near Langeleben, east of Brunswick, the last elements of Panzer-Abteilung 2106 were disbanded. Other parts of the brigade had been operating in small Kampfgruppen formed from the surviving elements of the Panzer and Panzergrenadier battalions and in early April these were trapped in the Ruhr Pocket.

Panzer-Abteilung 2107. Commanded by Major Hans-Albrecht von Plüskow, the battalion was created on 28 July 1944 from elements of Panzer-Abteilung 125, Reserve-Panzer-Abteilung 8, together with drafts from various training and replacement establishments of Wehrkreis V, and attached to Panzer-Brigade 107.

On Monday, 4 September 1944 OKW advised that the brigade's training period was at an end and while this was not unusual for the independent Panzer brigades it meant that most of the tank

Another view of the Panzer IV/70(A) shown on the previous page. As the application of Zimmerit was discontinued in the second week of September 1944, as few as sixty or seventy of these vehicles may have received a coat of the anti-magnetic mine paste.

Abandoned in the main street of Ober Bessenbach, a village less than 5 kilometres south-west of Aschaffenburg, this Panzer IV/70(A) tank destroyer was disabled by US Army units just hours before the photograph was taken on 1 April 1945. Note the lengths of spare track held by large hooks on the front of the superstructure. Although it is often associated with Panzer-Regiment 2, when it was attached to Panzer-Brigade 106, another possible candidate is Heeres-Sturmgeschütz Brigade 280 which was attached to LXXXII.Armeekorps and fought in this area.

crews had been given a week or less to familiarise themselves with their vehicles. In fact, the Panzer IV/70(V) tank destroyers that were to equip the Panzer battalion's 4.Kompanie were not shipped until the following Friday.

Diverted from Belgium to Holland, the brigade went into action for the first time on 19 September 1944 when the tanks of Panzer-Abteilung 2107 supported Fallschirmjäger-Regiment 21 in an attempt to capture the bridge over the Wilhelmina Canal at Son, north of Eindhoven, while parts of the brigade were still being unloaded at Venlo. The brigade remained in this area fighting against the US Army paratroopers and British tanks that were attempting to drive north as part of Operation Market-Garden until 24 September 1944 when the remaining elements were withdrawn to the area around Overloon and Oploo to defend the Maas River crossing at Boxmeer. Faced by the overwhelming strength of a full American armoured division, the Germans abandoned Oploo and although Allied attacks at Overloon and Venlo were halted Panzer-Brigade 107 had suffered heavy losses.

By the end of October 1944 the brigade had been reduced to a weak Kampfgruppe with just eleven Panthers and eight Jagdpanzers on hand (1) and on 8 November the remnants were withdrawn to Truppenübungsplatz Baumholder, north-west of Kaiserslautern, and disbanded on the following day. The surviving elements were absorbed by the rebuilt 25.Panzergrenadier-Division.

Panzer-Abteilung 2108. Attached to Panzer-Brigade 108, the battalion had received its full allocation of Panther tanks by 18 September 1944 when the brigade started moving to the front but the eleven Panzer IV/70(V) tank destroyers authorised for 4.Kompanie were not shipped until the next day and they are not mentioned in the brigade's reports until the first week of October, by which time the Panzer battalion had been in combat for two weeks.

On 4 October 1944 the brigade reported that twenty-three Panthers were on hand although just twelve of these were considered combat-ready and that a total of eight Panzer IV/70(V) tank destroyers were with 4.Kompanie which suggests that the company had joined the brigade some days prior to this date as three vehicles had already been lost.

The brigade was formally disbanded on 30 October 1944 after the fighting for Aachen and the surviving tank destroyers and their crews were allocated to Panzerjäger-Abteilung 228 (see below).

Panzer-Abteilung Putlos. Sometimes referred to as Panzer-Lehr-Abteilung Putlos, this unit was raised from the staff and instructors of the army's tank gunnery school. The battalion was attached to Kampfgruppe Bennigsen, part of Panzer-Division Clausewitz, on 17 April 1945 and contained a single Jagdpanzer IV tank destroyer taken over from the Ersatzheer. Within days most of the Kampfgruppe was lost in the fighting near Fallersleben, north-east of Brunswick.

Notes

1. The true figure may have been significantly less as there was a tendency to record the brigade's tank destroyers and Flakpanzers as simply 'Pz.IV' without differentiating between the two.

Each division of the German army contained an organic anti-tank battalion referred to as a Panzerjäger-Abteilung (1). From the beginning of the war every effort was made to fully equip these battalions with enough motor transport to move their towed guns and the Panzergrenadier and Panzer divisions were increasingly supplied with self-propelled weapons. By 1944 these included a rather motley collection of anti-tank guns mounted on captured French and Czech tank chassis, some armed with captured Russian guns, and it was envisaged that the Jagdpanzer series of vehicles would eventually replace them. The Panzerjäger battalions that were allocated tank destroyers and that served in the West are listed below.

Panzerjäger-Abteilung 3. Attached to 3.Panzergrenadier-Division, the battalion was formed in June 1944 with one anti-aircraft (2) and two assault gun companies. In April 1944 thirty-one Jagdpanzer IV tank destroyers were allocated by the Heereszeugamt but these were not received by the battalion until late August. It is known that the first and second companies were each equipped with fourteen of these vehicles and it would seem reasonable to assume that the battalion headquarters acquired the remaining three.

The division took part in the fighting for Aachen in November 1944 and in December, Panzerjäger-Abteilung 3 reported that seventeen Panzer IV/70(V) tank destroyers were on hand, although by the end of the same month the number of original Jagdpanzer IV had been reduced to seven. These vehicles were split between the first and second companies.

The division, as part of 15.Armee, defended Cologne but was eventually trapped in the Ruhr Pocket where the survivors surrendered to units of the US Army in April 1945.

Panzerjäger-Abteilung 33. The battalion was attached to 15.Panzergrenadier-Division which served in Italy until September 1944 when it was transferred to eastern France and initially subordinated to 1.Armee in Lorraine. The battalion reported that twenty-nine Jagdpanzer IV were on hand at that time and presumably a number of these had come from Italy and six were shipped from the Heereszeugamt in August 1944 while the division was moving to France. In the same month six Sturmgeschütz III assault guns were received from Panzer-Brigade 113 which had been disbanded after the fighting in Lorraine.

In December 1944, when the division took part in the Ardennes Offensive, the number of assault guns had been reduced to two and in January 1945 just twelve Jagdpanzer IV were on hand. Later that month the remaining Sturmgeschütz III assault guns were handed over to the division's Panzer-Abteilung 115.

Panzerjäger-Abteilung 38. Attached to 2.Panzer-Division, the battalion received twenty-one Jagdpanzer IV tank destroyers

...........*text continued on page 13*

Notes

1. The employment of towed anti-tank weapons within infantry formations is somewhat more complicated but need not be examined here.
2. At least one authoritative source states that the third company was equipped with Pak 40 towed anti-tank guns.

This Jagdpanzer IV/70(V) from 2.Kompanie, 3.Panzergrenadier-Division, was photographed in company with the vehicle depicted at the bottom of page 23 of the Camouflage & Markings section of this book. Both Jagdpanzers had three return rollers and were fitted with the Flammenvernicter exhaust mufflers but were unusual in lacking any steel roadwheels. This suggests that they were assembled in early September 1944. The contrast between the three colours of the factory-applied camouflage scheme is quite clear here.

PANZERJÄGER-ABTEILUNG, JUNE 1944

From March 1944, Jagdpanzer IV L/48 tank destroyers were allocated to the Panzerjäger battalions of Panzer and Panzergrenadier divisions and it was planned that each battalion would be made up of two fully armoured companies and one company of towed anti-tank guns. The Panzerjäger battalions of Panzer divisions were organised according the KstN 1149 ausf A, which allowed for two companies each made up of ten vehicles with one additional Jagdpanzer for the battalion headquarters, giving a total of twenty-one tank destroyers. The anti-tank battalions of Panzergrenadier formations and the Panzer-Lehr-Division were allocated thirty-one Jagdpanzers and were built using KstN 1149 ausf B which authorised two fourteen-vehicle companies and a further three Jagdpanzers with the battalion headquarters.

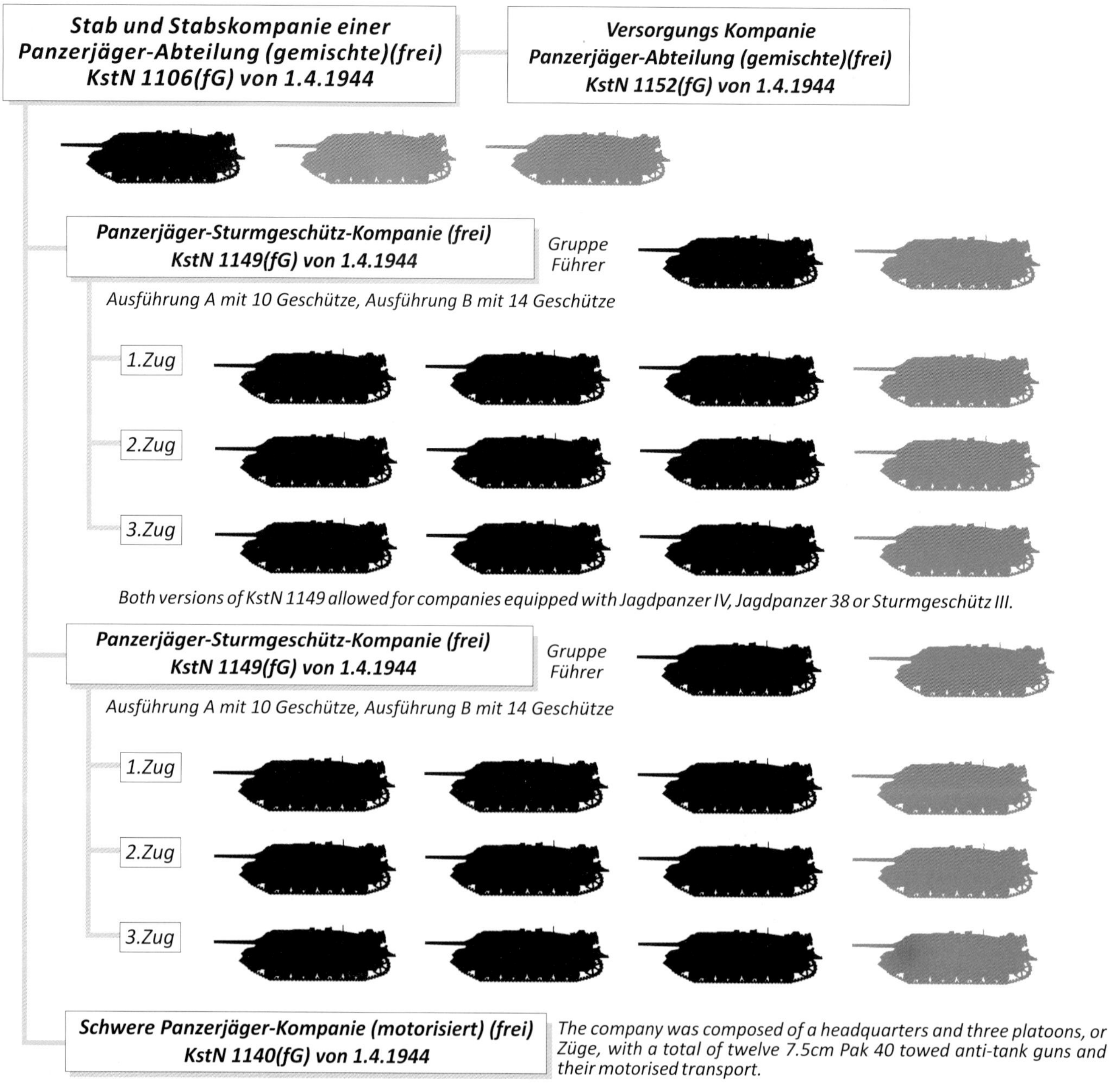

In November 1944 a new organisation was introduced to replace the company structure which was still made up of three platoons of either three or four tank destroyers but the company headquarters was reduced to one Jagdpanzer. Note that, officially at least, these companies could be composed of Jagdpanzer IV or Jagdpanzer 38 tank destroyers and the Sturmgeschütz III is no longer mentioned.

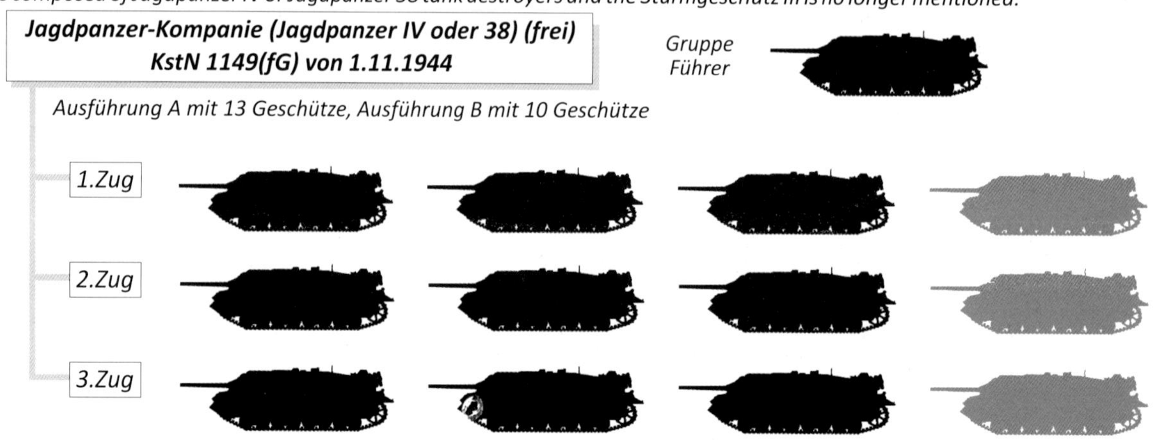

............*continued from page 11*

in April 1944 and initially these were distributed among the three companies as no towed anti-tank guns were available. In early June 1944, 3.Kompanie was issued with its authorised 7.5cm Pak 40 guns and its Jagdpanzer IV, possibly as many as eight, were handed over to the first and second companies. No replacements were allocated during the Normandy battles and by the end of August all but one of the Jagdpanzers had been lost.

The battalion was rebuilt in October and Novcmber 1944 and equipped with Sturmgeschütz III assault guns.

Panzerjäger-Abteilung 50. Attached to 9.Panzer-Division in early 1944, the battalion was made up of a staff element, two assault gun companies and a supply and maintenance company. The assault gun companies were equipped with a mixed bag of Sturmgeschütz III assault guns, Sturmhaubitze 42 and Marder III self-propelled guns.

In June 1944 the battalion was allocated twenty-one Jagdpanzer IV tank destroyers and although I can offer no definitive information on how these vehicles were distributed, it seems that the tank destroyers were all consolidated into a third assault gun company (1).

The battalion was rebuilt after the battles along the Siegfried Line and was allocated four Panzer IV/70(V) tank destroyers on 28 November 1944 and a further ten on 14 December. In addition to these a number of Jagdpanzer IV had survived the Normandy battles. On 15 March 1945 the division reported that just eight Panzer IV/70(V) and two Sturmgeschütz III were on hand and it is likely that the latter were with Panzer-Regiment 33. See also Panzer-Regiment 33 above.

Panzerjäger-Abteilung 61. In June 1944 the battalion contained two assault gun companies and one company of towed anti-tank guns and was attached to 11.Panzer-Division. The battalion was allocated twenty-one Jagdpanzer IV in June but these may not have all arrived before August.

In November 1944 another twenty Jagdpanzer IV tank destroyers were shipped from the Heereszeugamt to replace those lost in the fighting in southern France and the subsequent retreat and this may have been the last shipment of Jagdpanzer IV vehicles made in the war. At this time the battalion reported that eight assault guns were on hand and this total probably included a number of Sturmgeschütz III ausf F/8 models.

The division took part in the Ardennes Offensive, reporting in December 1944 that fourteen Jagdpanzer IV were still on hand. In 1945 the division was involved in the attempts to destroy the American bridgehead at Remagen and subsequently retreated towards the south-east, surrendering to US Army units near the town of Passau on the Austrian border

Notes

1. The battalion's 3.Kompanie is mentioned by Tim Haasler in his book *Hold the Westwall: The History of Panzer Brigade 105, September 1944* but I feel that the creation of a third company may have been a temporary expedient. The Marder self-propelled guns are not recorded as being on hand after August 1944 and in December the battalion reported that just fourteen Sturmgeschütz III were available.

The official caption to this photograph gives the location as St Jean Rohrbach but the road sign in the background confirms that it is in fact Püttlingen, a village some 20 kilometres to the north, across the German frontier, between Saarbrücken and Saarlouis. Although difficult to see here, a very faint rendering of the 11.Panzer-Division unit insignia can be discerned in the original print and this is shown more clearly in our illustration on page 20.

Captured intact during the Normandy battles, this Jagdpanzer IV L/48 tank destroyer has been associated with a number of different units but was almost certainly allocated to 3.Kompanie, Panzerjäger-Lehr-Abteilung 130. Other photographs in this series show that this vehicle was fitted with a second machine-gun aperture on the hull glacis above the driver's visor which was dropped from production in March 1944 and initially replaced with a circular plate. Note the rack on the hull rear holding lengths of spare track.

on 2 May 1945. Photographic evidence confirms that a number of the battalion's Jagdpanzer IV were still operational at that time.

Panzerjäger-Lehr-Abteilung 130. Formed in April 1944 from elements of III.Abteilung, Panzer-Lehr-Regiment, the battalion was attached to the newly created Panzer-Lehr-Division (1).

Shortly after its formation the battalion received thirty-one Jagdpanzer IV tank destroyers and initially the first and second companies were each assigned fourteen vehicles while the headquarters had three. By the beginning of June 1944 this arrangement had been changed and the battalion was made up of three mixed companies each with nine Jagdpanzer IV and four towed 7.5cm Pak 40 guns. The remaining four tank destroyers were with the battalion headquarters.

The battalion received twenty-one Panzer IV/70(V) tank destroyers in mid-November 1944 in time to take part in the defence of the Saar industrial region where they were able to repulse a strong American armoured thrust. On 14 December 1944, just two days prior to the commencement of Operation Wacht am Rhein, the battalion was allocated a further eleven Panzer IV/70(V) vehicles. The figure of fourteen tank destroyers on hand given in some sources for this time almost certainly represents fully operational vehicles. In the early stages of the Ardennes Offensive the battalion's 3.Kompanie, with nine Pak 40 anti-tank guns, formed part of a Kampfgruppe led by Major Gerd von Fallois, the commander of the division's reconnaissance battalion, while the other companies were held in reserve (2).

In early January 1945 the division was temporarily withdrawn from the front line and a report from this time mentions a total of nine Jagdpanzer IV on hand although this entry almost certainly refers to the remaining Panzer IV/70(V) tank destroyers as any of the former that had survived the fighting in France were handed over to Panzerjäger-Abteilung 200 in October 1944 (see below). In March the division took part in the fighting in Holland along the Maas Line and in the same month was involved in the attempts to destroy the American bridgehead at Remagen as part of a Kampfgruppe made up from the remnants of 9.Panzer-Divison and 11.Panzer-Division. After the failure of this operation the division was trapped in the Ruhr Pocket and surrendered to units of the US Army on 15 April 1945.

Panzerjäger-Abteilung 200. Originally raised in June 1943, the battalion was attached to the newly-rebuilt 21.Panzer-Division. The battalion was initially equipped with towed 8.8cm Pak 43 anti-tank guns but most of these were lost in the Normandy battles and the retreat through France and in October 1944 the battalion was reformed. Comprising a staff element, two assault gun companies and a

Notes

1. The Panzer-Lehr-Regiment was formed in 1938 as a training unit and by the autumn of 1940 consisted of three battalions. Regiments and battalions raised for the Panzer-Lehr-Division were all numbered as 130 to differentiate them from the various training and replacement formations.
2. The authorised strength of 3.Kompanie is given as twelve towed Pak 40 anti-tank guns so it would seem that the tank destroyers were by this time concentrated in the first and second companies and probably the headquarters. This reorganisation may have taken place when the Panzer IV/70(V) tank destroyers began arriving in November.

towed anti-tank gun company the battalion was supposed to be equipped with assault guns or tank destroyers but these were in such short supply that the first and second companies were allocated a number of self-propelled guns based on captured French Hotchkiss tanks.

In addition, a number of Jagdpanzer IV, as many as three, were taken over from Panzerjäger-Lehr-Abteilung 130. The third company received an allocation of new 8.8cm Pak 43 guns and a platoon of Russian 85/88mm Flak guns which were later replaced by 7.5cm Pak 40 guns.

In November the Hotchkiss self-propelled guns were relinquished and in late December 1944, seventeen Panzer IV/70(V) tank destroyers were shipped to the battalion and these arrived in the first weeks of 1945. The tank destroyers, including the single remaining Jagdpanzer IV, were operated by the first and second companies while the third company retained its towed guns (1).

In February 1945 the battalion, with its parent formation, was transferred to the Eastern Front and later disbanded.

Panzerjäger-Abteilung 228. The battalion was attached to 116.Panzer-Division and took part in the Normandy campaign equipped with twenty-one Jagdpanzer IV that had been received in July 1944. It is unclear how many of these survived the battles of July and August but no other replacements were allocated until September 1944 when ten Panzer IV/70(V) tank destroyers were issued by the Heereszeugamt.

A report prepared by the division for the commander of Panzertruppen West dated 9 November shows a total of twenty vehicles listed under 'Panz Jäg.IV' with ten described as 'le.', or light, and the remaining ten as 'schw.', or heavy. I can offer no definitive explanation as to why the vehicles should be described in this way but it may have been a convenient method of distinguishing the Panzer IV/70(V), with its longer gun, as heavy. If correct this would mean that ten Jagdpanzer IV survived the Normandy battles and the subsequent retreat. Jeff Dugdale, the author of several works on the organisations of the German Army and Waffen-SS, states that just two weeks prior to the commencement of the Ardennes Offensive the battalion was organised with a headquarters made up of one Jagdpanzer IV, one Sturmgeschütz III and a Sdkfz 251 halftrack while 1.Kompanie contained six Sturmgeschütz III assault guns. The second company was equipped with nine Jagdpanzer IV while the third had eight Panzer IV/70 tank destroyers.

On 14 December 1944, five Panzer IV/70(V) vehicles were despatched by the Heereszeugamt and these were the last that the battalion would receive (2).

Notes

1. As a matter of interest, 2.Kompanie also operated a Sturmgeschütz III assault gun in February and March 1945 while 1.Kompanie acquired a Marder II self-propelled gun at the same time.
2. A 9 December 1944 allocation of six Panzer IV/70 (V) is mentioned in some accounts but it seems that the vehicles never arrived.

Photographed after the fighting around Mont Ormel, the final phase of the battle for the Falaise Pocket, this early production Jagdpanzer IV L/48 is from 2.Kompanie, Panzerjäger-Abteilung 228. Note the large stowage box fixed to the rear hull. This vehicle is unusual in retaining the muzzle brake for the 7.5cm gun as most of these were removed by front-line units in April 1944. This vehicle is also shown in the Camouflage & Markings section on page 18.

The Panzer IV/70(V) tank destroyer shown in this photograph has the waterproof cover that was issued for these vehicles and the Jagdpanzer IV L/48 from September 1944, held in place by a number of small hooks welded to the superstructure sides, front and rear. Note the later-style gun support which lacks the cut-out of the earlier versions and the covers for the brake access hatches on the hull front without air-intakes, indicating that this vehicle was assembled in February 1945 or later. This image is taken from a series which shows the survivors of 11.Panzer-Division making their way towards the American lines between Passau and Wittenberg in May 1945 after the German surrender. However, the division was never allocated any of these Jagdpanzers and this vehicle may be from a Kampfgruppe made up from elements of 2.SS-Panzer-Division, which had fought their way through Austria towards the West, or Panzerjäger-Lehr-Abteilung 130 which was also in the area.

This late production Panzer IV/70(V) tank destroyer has also been associated with 11.Panzer-Division in the past but, as mentioned above, the division never received any of these vehicles. Note the late pattern gun support, the brake access hatches without air-intakes and the lack of an armour panel on the rear hull which all point to this vehicle being assembled in February 1945 at the earliest. This may be one of the last ten Panzer IV/70(V) tank destroyers produced by Vomag in April 1945 which were picked up from the factory by crews of II.Abteilung, Panzer-Regiment 33. The 5.Kompanie number would also seem to confirm this identification. This vehicle is also shown on page 24 of the Camouflage & Markings section of this book.

2

1

5

1. Jagdpanzer IV ausf F L/48. 2.Kompanie, SS-Panzerjäger-Abteilung 17. France, July 1944. Photographed near Vaiges shortly after the battalion was engaged at Laval, this vehicle is one of the thirty-one tank destroyers allocated to the battalion in June 1944. The distinctive unit insignia of 17.SS-Panzergrenadier-Division (2) is shown here as a matter of interest only and there is no evidence that it was applied to the Jagdpanzers.

3. Jagdpanzer IV ausf F L/48. 3.Kompanie, Panzerjäger-Abteilung 228. France, July 1944. Attached to 116.Panzer-Division, the battalion had twenty-one of these tank destroyers on hand during the Normandy battles. The company number, depicted on the hull side, is largely conjectural as most of these vehicles were covered with foliage camouflage. The division's unit insignia and the tactical symbol denoting a Jagdpanzer IV company with the company number (4) were carried on the hull front and these are clearly visible in our photograph (5).

3

4

1

3

2
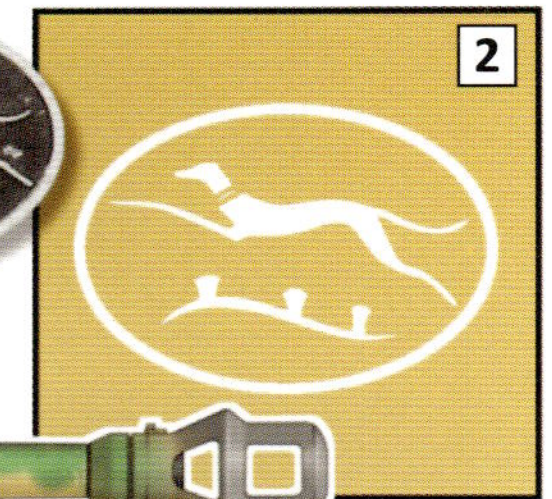

5. It is usually suggested that patches of Zimmerit were removed to accommodate the company number of the Jagdpanzer depicted below but in fact an extra layer of paste has been added, as can be seen from our photograph.

5

1. Jagdpanzer IV ausf F L/48. 2.Kompanie, Panzerjäger-Abteilung 228. France, August 1944. Photographed after the fighting at Mont Ormel, east of Falaise, this Jagdpanzer is unusual in that it retains the muzzle brake for the 7.5cm gun. Attached to 116.Panzer-Division many tank destroyers carried the division's unit insignia (2) which was also worn as a semi-official cap badge (3). Note the large box on the hull rear. A photograph of this vehicle is reproduced on page 15.

4. Jagdpanzer IV ausf F L/48. Unit uncertain. France, summer 1944. This is a very early production vehicle identified by the armoured cover for the machine gun aperture on the left-hand side. Although this tank destroyer is sometimes associated with SS-Panzerjäger-Abteilung 12 it was almost certainly operated by 3.Kompanie, Panzerjäger-Lehr-Abteilung 130.

4

1

2

1. Jagdpanzer IV ausf F L/48. SS-Panzerjäger-Abteilung 9. Holland, autumn 1944. This battalion commonly covered their Jagdpanzer tank destroyers in a thick layer of concrete, avoiding the machine-gun aperture and driver's visor. Our photograph (2) shows that the hull front was also sometimes covered.

3. Jagdpanzer IV ausf F L/48. Unit uncertain. Western Front, probably early 1945. This early production vehicle, evidenced by the conical-shaped armoured cover next to the main gun, is painted in what seems to be an imitation of one of the factory-applied camouflage schemes that were introduced in mid-August 1944. Note the solid black Balkenkreuz and the name Gisela (4) painted on the superstructure side.

3

4

1. Jagdpanzer IV ausf F L/48. 2. Panzerjäger-Abteilung 61. Germany, late 1944. Photographed near Püttlingen, after the battles for Metz, this vehicle is painted in the so-called ambush camouflage scheme which was applied at the factory. Large patches of Olivgrün and Rotbraun were painted over the Dunkelgelb base coat, covering about two-thirds of the vehicle. A stencil, made up of the negative shapes created by a design of large overlapping circles, was used to apply Dunkelgelb and one of the dark shades, shown here as Olivgrün, in a random pattern. Lastly, patches of Dunkelgelb were sprayed over the vehicle. In the original print, a copy of which appears on page 13, the unit insignia of 11.Panzer-Division (2) is just visible on the hull front (3).

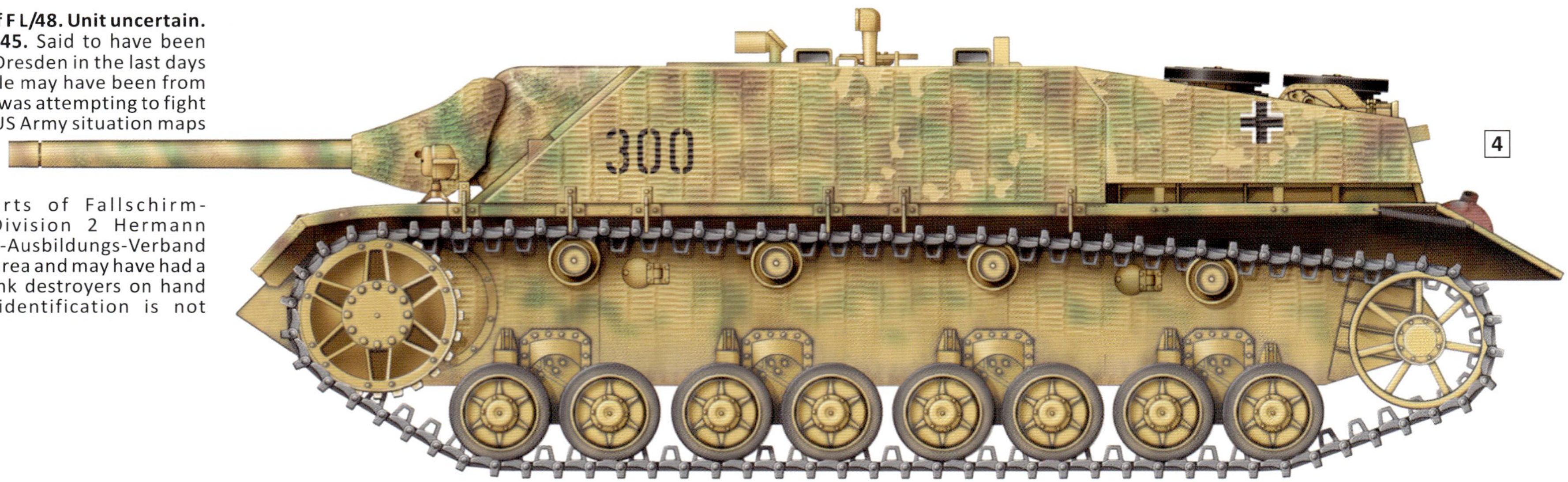

4. Jagdpanzer IV ausf F L/48. Unit uncertain. Germany, March 1945. Said to have been photographed near Dresden in the last days of the war this vehicle may have been from one of the units that was attempting to fight its way to the West. US Army situation maps indicate that parts of Fallschirm-Panzergrenadier-Division 2 Hermann Göring and Panzer-Ausbildungs-Verband Franken were in the area and may have had a number of these tank destroyers on hand but any definite identification is not possible.

1

2

3

1. Panzer IV/70(V). Panzer-Abteilung 2107. Holland, autumn 1944. Attached to Panzer-Brigade 107, the battalion was rushed to the front after a short period of training and this early production vehicle is devoid of any camouflage or markings other than the standard Balkenkreuz. Gun barrels were coated in a heat-resistant laquer and it was common to see unpainted examples in the field. A detail from one of our available photographs (2) shows how roughly the Zimmerit was applied to the mantlet. The unit insignia of Panzer-Brigade 106 (3) is shown here as a matter of interest only and it is unlikely that it was carried by any of the battalion's tank destroyers. Photographs of this vehicle are reproduced on pages 8 and 9.

4

4. Panzer IV/70(V). Unit unknown. Western Front, late 1944 or early 1945. Note the curved Absatzkrümmer attachment on the exhaust mufflers identifying this as a very late production vehicle. I have not been able to discover exactly when these were fitted to the Panzer IV/70(V) but they were not incorporated into Panther production until late January 1945. Note that the camouflage, which would seem to be a factory-applied scheme, extends to the roadwheels but not the lower hull. A photograph of this vehicle appears on page 50.

3

1. Panzer IV/70(V). Unit uncertain. Germany, 1945. Photographed at some time after the end of the war, the only thing known for certain about this vehicle is that it is pictured on the Western Front. Production details, which may have given an approximate date of assembly, are simply missing. What may be the unit insignia of the Führer-Grenadier-Brigade (2) can be seen in our photograph (3) to the right of the driver's visor but the brigade's Jagdpanzers were all supposed to have been concentrated in 4.Kompanie of the Panzer battalion. The style of numbering is very similar to that seen on the vehicle depicted at the bottom of page 21.

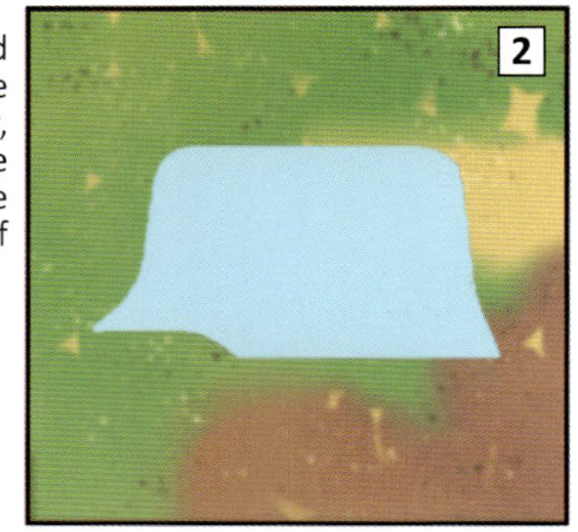

2

1

5

4. Panzer IV/70(V). 2.Kompanie, Panzerjäger-Abteilung 3. Germany, early 1945. Attached to 3.Panzergrenadier-Division, the battalion took part in the battles for the Remagen bridgehead in March 1945 and this tank destroyer was abandoned on the northern edge of the front near Oberpleis. The very neat factory-applied camouflage scheme extends to the lower hull and probably the roadwheels, although the colours of the latter are somewhat speculative on my part. Note the distinctive Sternantenna D on the rear hull identifying this tank destroyer as a command vehicle. The unit insignia of 3.Panzergrenadier-Division is shown at right (5). It is not known if the division's Jagdpanzers carried this marking but some of the Sturmgeschütz assault guns certainly did.

4

The vehicles depicted on this page were photographed at Oldenburg, west of Bremen in Northern Germany, when the battalion surrendered in May 1945.

1

2

1. Panzer IV/70(V). Schwere Panzerjäger-Abteilung 655. Germany, spring 1945. The tank destroyer shown here is a late production version fitted with the Absatzkrümmer exhaust muffler extensions and the later-style solid gun support but does not have the armoured panels on the hull side behind the main superstructure. The latter were dropped from production at some time in early 1945, possibly March. These features can also be seen in our photograph (2). Note that both vehicles are fitted with the early pattern rear idler. Note also the eyes and mouth painted onto the mantlet of the nearest vehicle.

3

3. Panzer IV/70(V). Schwere Panzerjäger-Abteilung 655. Germany, spring 1945. Part of this vehicle can also be seen in our photograph (2) above. Note the Sternantenna D wireless aerial for the FuG8 radio, the single steel road wheel and the two Balkenkeuz national markings. Note also the cast rear idler suggesting that this tank destroyer was assembled in February or March 1945. The photograph on which this illustration is based shows that this vehicle was built with the later brake access hatches which lacked an air-intake, further confirming the production date. Note the small step welded to angled front of the hull.

1

1. Panzer IV/70(V). Unit unknown. Western Front, late 1944 or early 1945. Although the photograph on which this illustration is based lacks any annotation or date, the presence of a US Army M9 12-wheel trailer would at least seem to confirm the location as the Western Front. What may be Pilze sockets are visible in our photograph and this would give a production date of November 1944 or later. The unusual hull Schürzen are almost certainly a field modification and great care has been taken to apply the camouflage. A very similar arrangement was used by Panzer-Regiment 20, which served on the Eastern Front, where the plates on the left-hand side were hinged to open upwards allowing access to the filler caps.

3

1. Panzer IV/70(V). Possibly 5.Kompanie, Panzer-Regiment 33. Germany, spring 1945. Also shown in our accompanying photograph (3) this vehicle is fitted with the solid gun support, later brake access hatches on the hull front, Flammenvernichter exhaust mufflers and fittings for the Entfernungs-Messer (EM) 0.9m rangefinder around the commander's hatch. In addition, there is no armoured panel on the rear hull side and this vehicle must be one of the last Panzer IV/70(V) tank destroyers to leave the assembly line. Note the armoured cover for the Sternantenna D aerial just visible in our photograph behind the Balkenkreuz. The identification with Panzer-Regiment 33 is discussed further on page 16 where a larger version of the same photograph is reproduced.

2

2

1

1. Panzer IV/70(A). Unit uncertain. Germany, spring 1945. Photographed near Tegersee, south of Munich near the Austrian border, this vehicle may be one of the last tank destroyers of Panzer-Brigade 106. This vehicle does not seem to have been fitted with the rails which held the hull Schürzen and the brackets for the gun support have been removed, as can be seen from our photograph (2). It also has four return rollers and the early bolted-on couplings at the hull front suggesting that this is one of the sixty tank destroyers assembled in September 1944.

4

1. Panzer IV/70(A). II.Abteilung, Panzer-Regiment 2. France, spring 1945. Captured after the fighting around Colmar in December 1944, this Panzer IV/70(A) has the full set of brackets to hold the Drahtgeflecht wire mesh panels in place but the rails have been shot away. Note the cast rear idler which was fitted to approximately half the Pzkpfw IV ausf J tanks on which these vehicles were based. Our photograph (4) shows that the Balkenkreuz was also carried on the rear of the superstructure and confirms that the roadwheels were painted in different colours. Another photograph of this vehicle is shown on page 4.

3

1

1. Panzer IV/70(A). Führer-Begleit-Brigade. Belgium, winter 1944. Photographed after the fighting in the Ardennes, this vehicle, like many others, has lost most of the rails and their supports that held the wire mesh panels shown in the illustration below. The brigade's Panzer battalion was made up in part from II.Abteilung, Panzer-Regiment Grossdeutschland and the thirty-eight Panzer IV/70(A) tank destroyers probably made up most of the three companies. As serious production only got underway in September 1944, most of these vehicles were finished in a factory-applied scheme as shown here. The company number is rendered here as red with a white outline but it is possible that the centre could have been some other dark shade. It is not known if the vehicles of the Panzer battalion carried the brigade's unit insignia (2) but the halftracks and unarmoured vehicles certainly used variations.

4

2

5

3

3. Panzer IV/70(A). Heeres-Sturmgeschütz-Brigade 341. Germany, March 1945. This late production vehicle is fitted with the Kugellafette Vorsatz P mount (4) in the loader's hatch, authorised for production in November 1944. Although a number were probably fitted from that month the curved barrel for the Sturmgewehr 44 was not available until January 1945. Note also the rack on the superstructure side and the stowage rails on the rear engine deck. These fittings are seen on vehicles photographed during the last weeks of the war, on both fronts, and may have been a late production modification that went undocumented. A photograph of this tank destroyer is reproduced on page 53. The brigade's unit insignia (5) is shown here as a matter of interest only.

JADGPANZER IV

AUSF L/48

1/35 SCALE

ANGEL IÑIGO

Angel's replica uses parts from Dragon Models' Jagdpanzer IV/70(V) kit with a replacement gun barrel from Voyager Model, which comes complete with a resin Topfblende mantlet. The images here and on the following page show the model in the first stages of completion as well as the largely scratch-built interior. Most of the photo-etched details are from Aber while the Schürzen are from Voyager Model. Several small parts, such as the ammunition box and helmets, are from the Tamiya Jagdpanzer L/48 model.

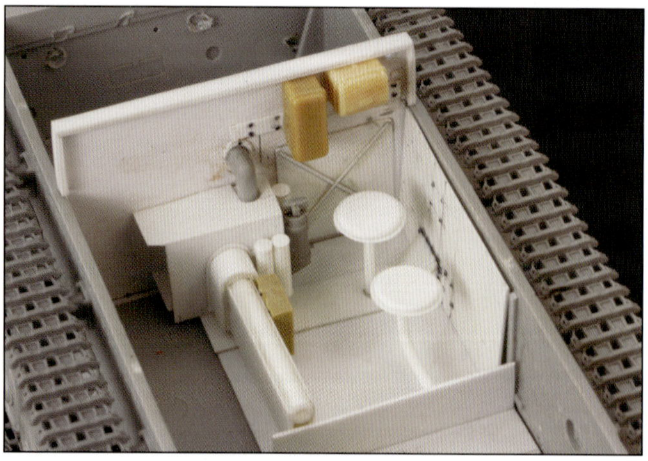

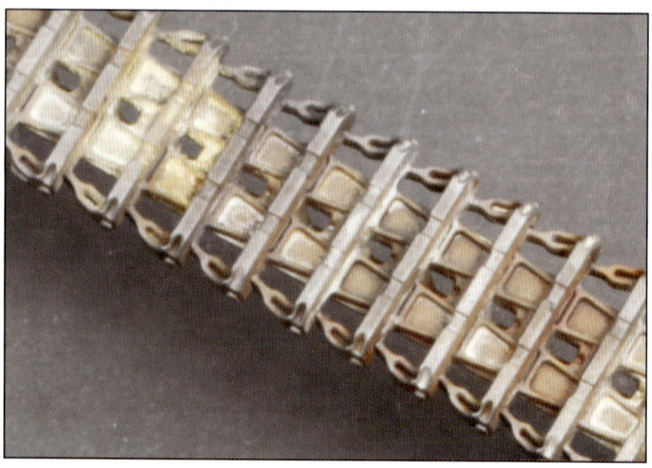

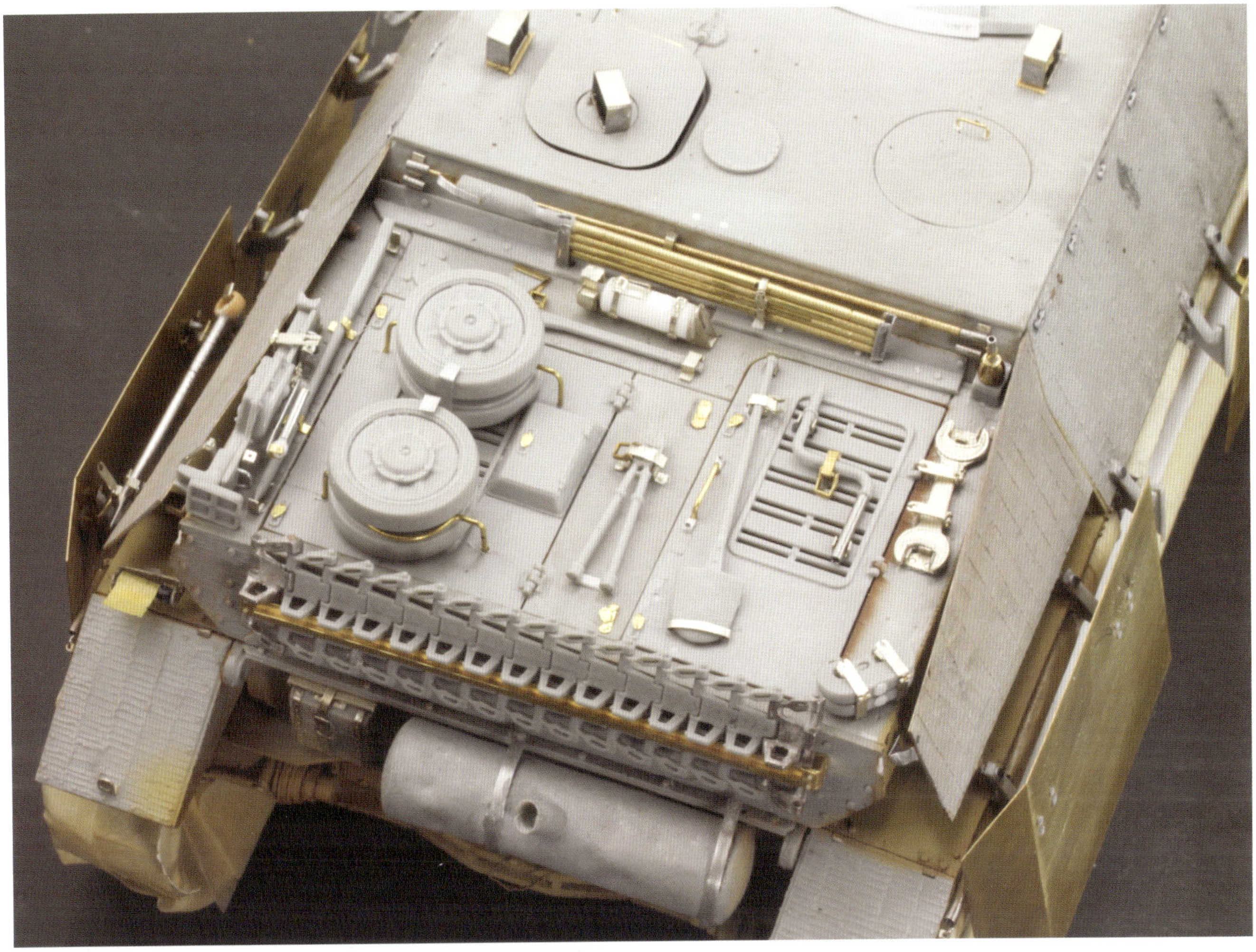

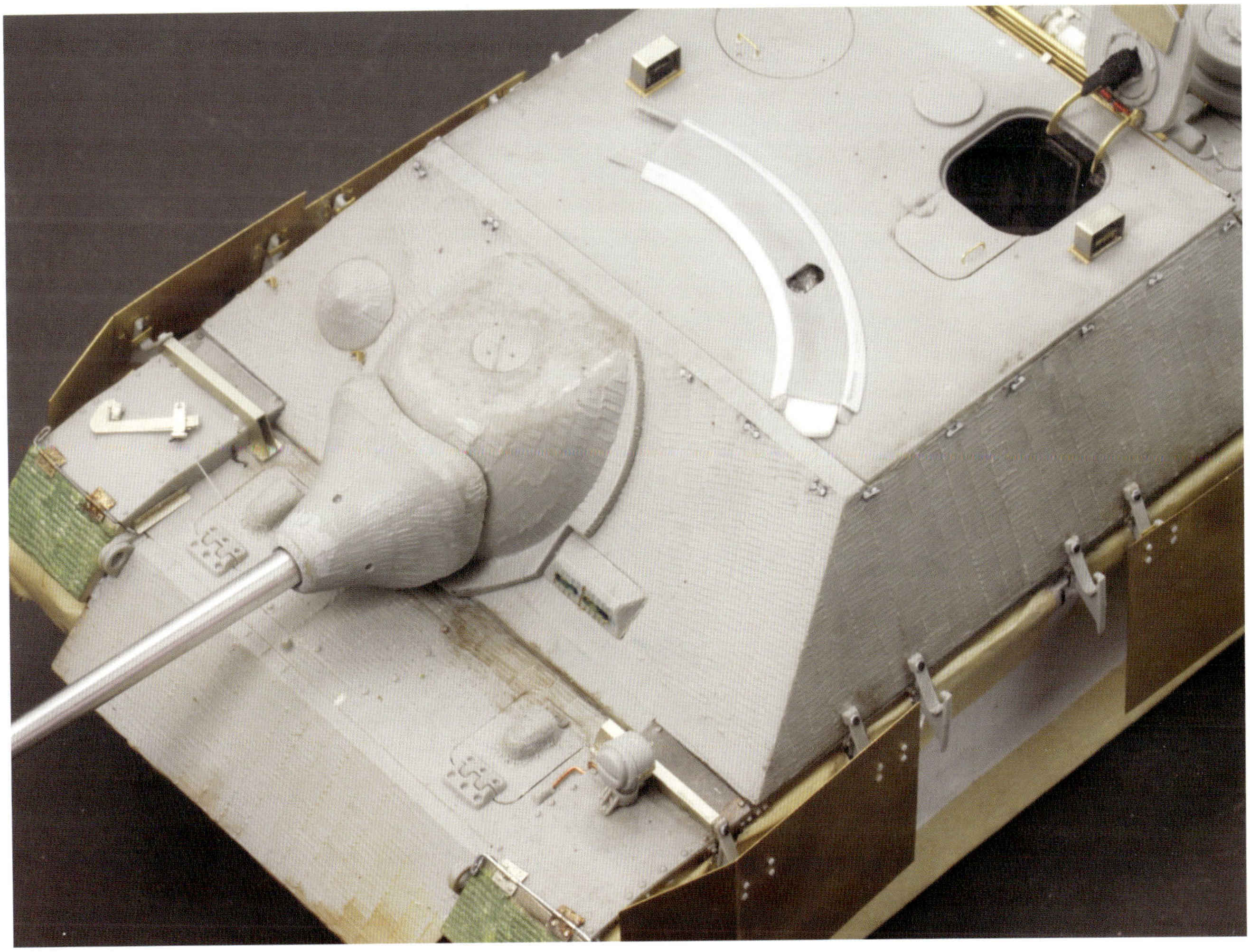

The assembled model went through fifteen separate stages of painting to achieve the effect seen here. Angel first primed and painted the vehicle with Tamiya acrylics and then added the weathering with Vallejo and Citadel paints and finished with Tamiya pigments and powdered graphite.

PANZER IV

L/70 (V) TANK DESTROYER
1/35 SCALE

PENNY SHI

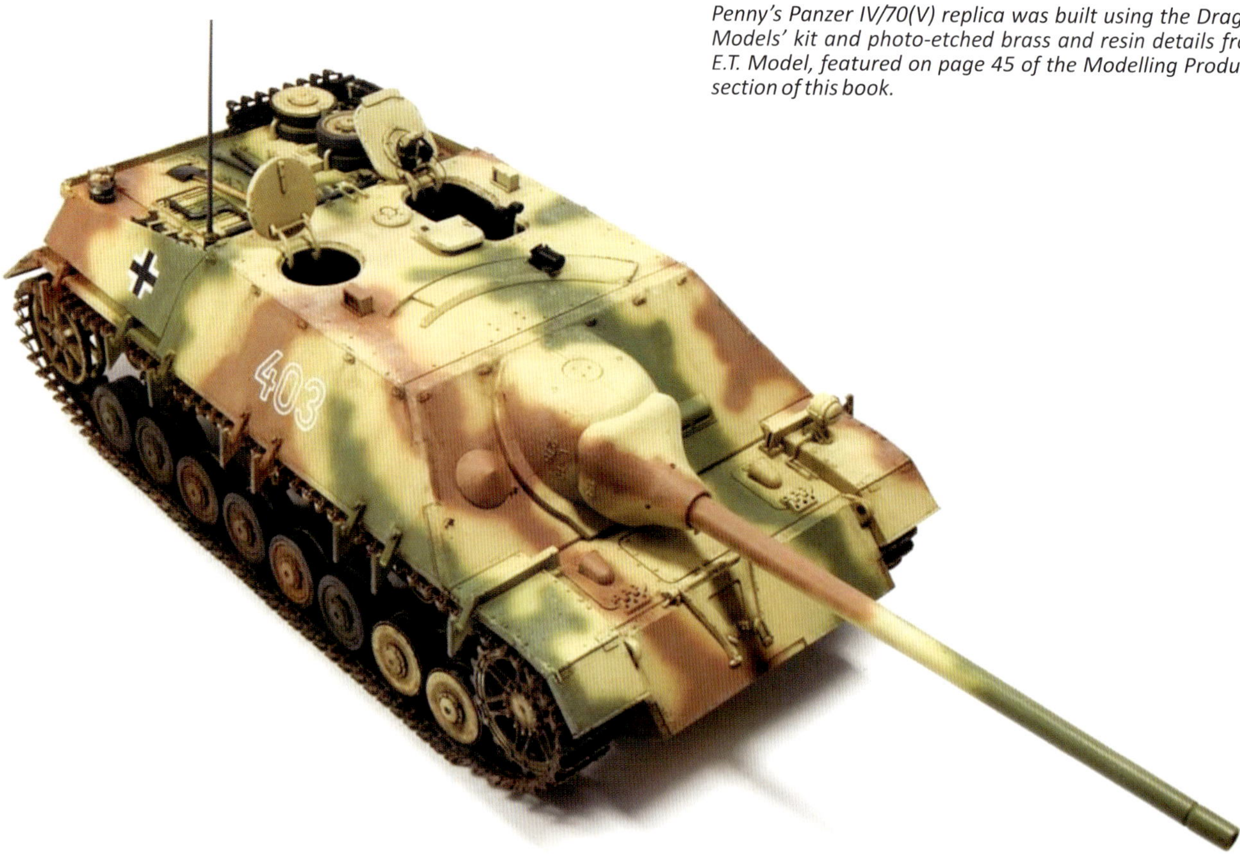

Penny's Panzer IV/70(V) replica was built using the Dragon Models' kit and photo-etched brass and resin details from E.T. Model, featured on page 45 of the Modelling Products section of this book.

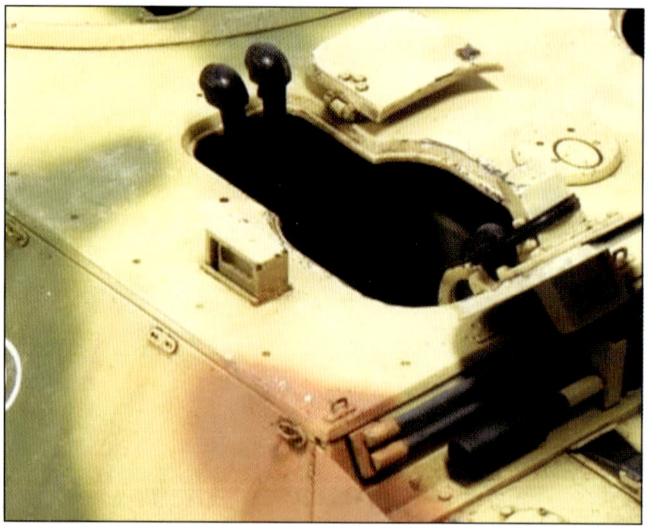

Added details include a resin and brass fire extinguisher, tools and clasps, Flammenvernichter exhaust mufflers and a complete set of fenders and track guards.

PANZER IV

L/70 (V) TANK DESTROYER

1/35 SCALE

KENTARO TAKABAYASHI

Kentaro's model is based on the 1997 Gunze Sangyo kit which was actually a re-boxing of Dragon Models' Jagdpanzer IV/70(V) Command Version model, released at almost the same time and with the same artwork by Vince Wai. Below is the assembled model built almost straight from the box. The images above show the stages of painting the Ambush camouflage pattern and the final weathering.

The completed model after painting. Kentaro used Tamiya acrylics and made a stencil to achieve the Ambush scheme. Small parts, such as the periscopes and tools, were painted by hand. The final weathering was completed with washes of watercolour paints and the paint chipping effect was created with a fine brush. The markings are those supplied with the kit and represent a vehicle of 116.Panzer-Division fighting in Germany in the spring of 1945.

PANZER IV
L/70 (V) TANK DESTROYER
1/35 SCALE
LIM KIAN GUAN

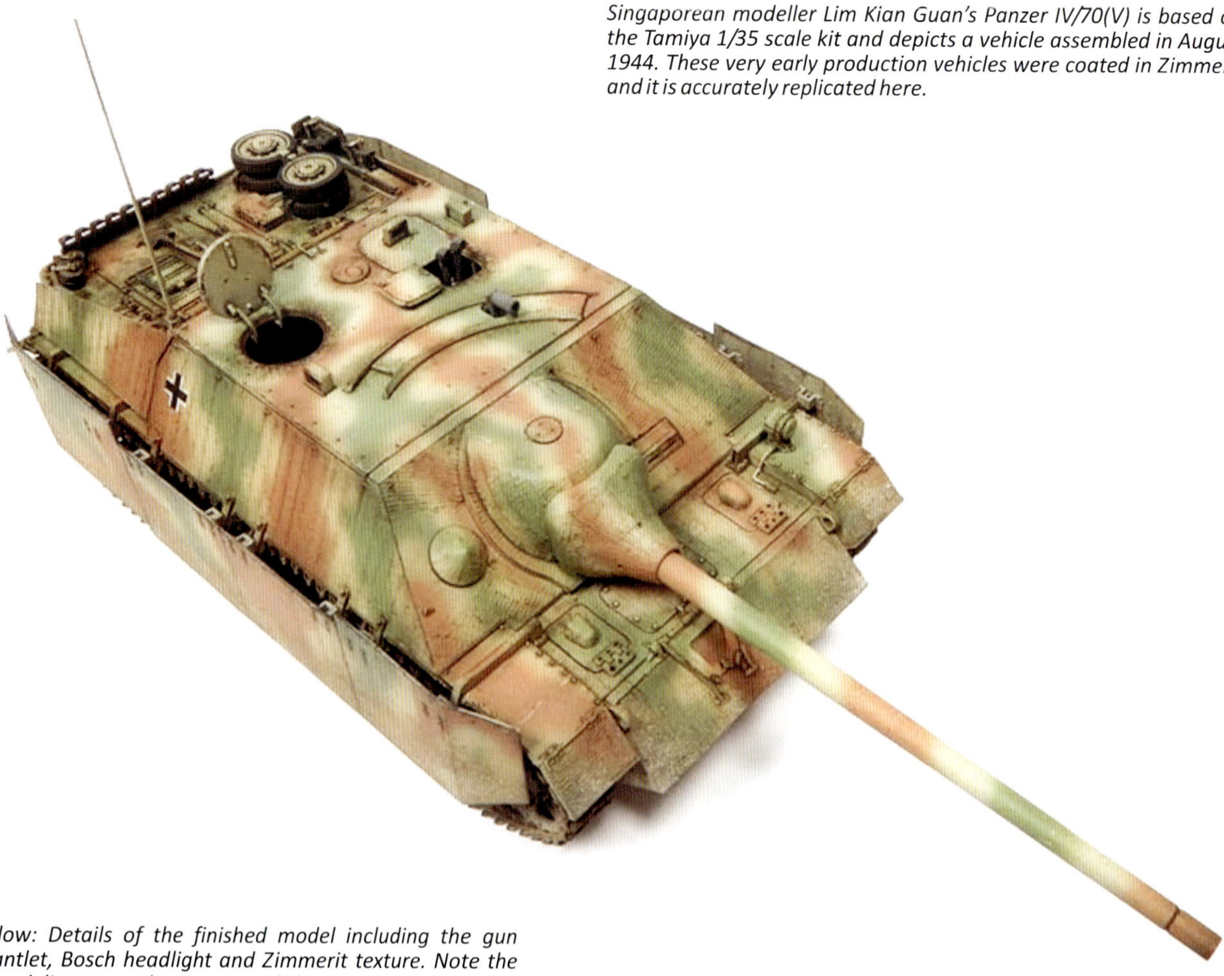

Singaporean modeller Lim Kian Guan's Panzer IV/70(V) is based on the Tamiya 1/35 scale kit and depicts a vehicle assembled in August 1944. These very early production vehicles were coated in Zimmerit and it is accurately replicated here.

Below: Details of the finished model including the gun mantlet, Bosch headlight and Zimmerit texture. Note the very delicate weathering around the crew hatches and on the engine deck.

PANZER IV

L/70 (V) TANK DESTROYER

1/76 SCALE

THEODOROS KALAMATAS

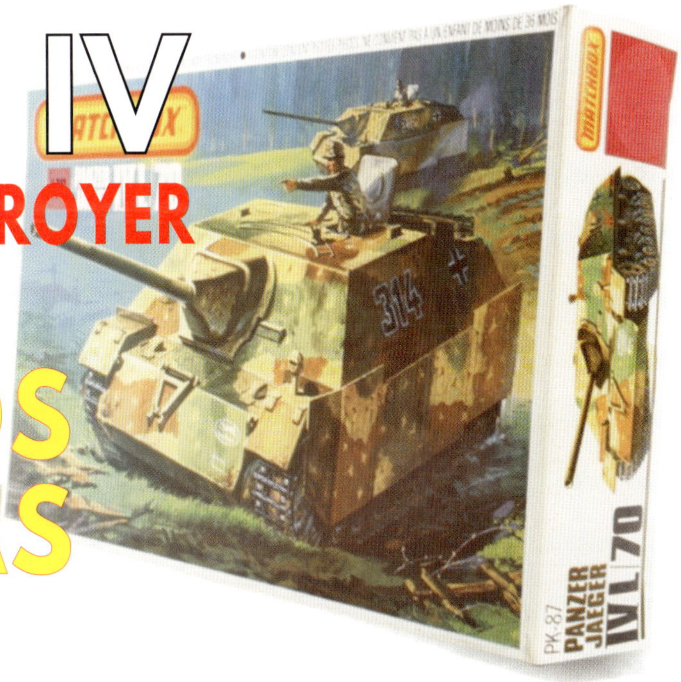

Theodoros' rendering of the 1/76 scale Matchbox kit, first released in 1978, was painted with Tamiya acrylics and weathered with a number of different products from Ammo by MIG and AK Interactive. The subtle depiction of the Ambush camouflage is particularly impressive, particularly in this small scale. The finished model was built straight from the box without any improvements or alterations.

Theodoros has managed to acquire the complete Matchbox 1/76 scale collection and expects to have them all built and painted to this standard by the end of 2020.

Although scale model releases of the Jagdpanzer IV series of vehicles are largely limited to offerings from the major manufacturers, the available kits are, for the most part, highly detailed and accurate replicas. They are produced in a number of scales from tiny 1/144 scale wargames miniatures to 1/16 scale motorised models. But the widest range by far are the 1/35 scale reproductions, which cover almost every factory modification, and it is those that I have chosen to concentrate on here.

The proliferation of models from the major manufacturers has seen the emergence of smaller, specialised companies producing additional items in resin, photo-etched metal and other materials allowing a level of super-detailing which is only limited by the modeller's skill and experience.

I have tried to include as many of these as possible but should point out that the following list is far from exhaustive, being more of a representative sample, and I would encourage readers to undertake their own research into the areas that interest them most. Similarly I have limited any comment or critique of these kits or products to those of which I have first-hand knowledge.

An index of manufacturers can be found on page 64.

TAMIYA INCORPORATED

This Japanese company, which began life as a sawmill, has been manufacturing plastic kits since 1959, releasing their first armour model two years later. By the early 1970s Tamiya was producing a large range of armoured vehicles and accessories and the company was almost solely responsible for the rise in popularity of 1/35 scale. Like Airfix, the adoption of a constant scale proved popular at a time when many manufacturers were making models to fit the box in which they would eventually be marketed. Tamiya's early kits were, for their day, highly detailed, relatively easy to assemble and, most importantly, affordable. From 2003 Tamiya began releasing a series of armour models in 1/48 scale and this line has been extremely successful for the company. At the time of writing Tamiya offers a 1/35 scale model of the Panzer IV/70(V) tank destroyer which was released in 2014 and is a complete re-tooling of their 1976 kit. The company also produces a 1/16 scale motorised model of the same vehicle complete with remote control, markings and crew figure.

At left: Tamiya's 1/16 fully motorised Panzer IV/70(V) model. Below: The 1/35 scale Panzer IV/70(V) built and painted as suggested by the kit instructions. The other marking option is shown on the front cover.

Below: Detail shots of the Tamiya 1/35 scale Panzer IV/70(V) shown at right including the crew figures, steel-rimmed road wheels and engine deck.

DRAGON MODELS LIMITED

At the time of writing this Hong Kong-based company's catalogue contained over twenty models of the Jagdpanzer series of tank destroyers in 1/35 and 1/72 scale including a model of the Jagdpanzer IV A-O mild steel prototype. It should be noted that a number of these are re-packaged kits, some containing additional parts or markings, and eleven are prepainted 1/72 scale display models. Having said that, the Dragon kits of the Jagdpanzer IV L/48, Panzer IV/70(V) and Panzer IV/70(A) are highly detailed replicas containing optional parts, photo-etched elements and a choice of markings. To complement the vehicles there is a large range of highly detailed, plastic multi-pose figures. Readers will note that, like most manufacturers, Dragon refers to all these vehicles as Jagdpanzer IV.

Above: Examples of the box art for Dragon Models' Jagdpanzer IV series of tank destroyers, the last two by the great Ron Volstad. At right and below: 1. Dragon's Panzer IV/70(V) Command Version November 1944 production kit in 1/35 scale. 2 and 3. Detail shots of the model shown at right. 4. The 1/35 scale Jagdpanzer IV A-0 mild steel prototype with the distinctive rounded hull front. 5. The Jagdpanzer L/48 model in 1/35 scale showing the early cylindrical exhaust muffler. 6 and 7. Dragon's Jagdpanzer IV/70(A) model, also in 1/35 scale. The detailed Thoma Schild mesh skirts are included in the kit.

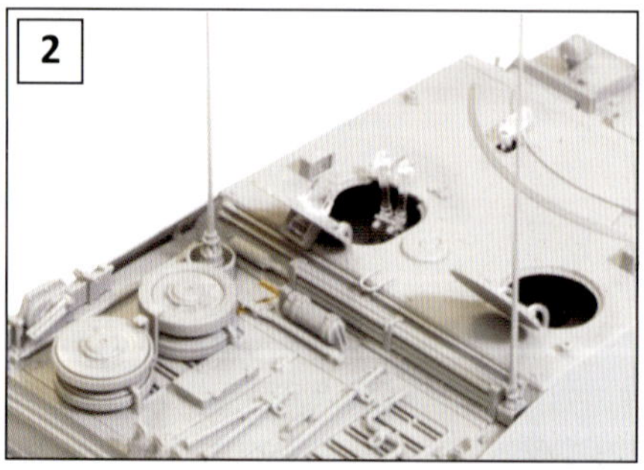

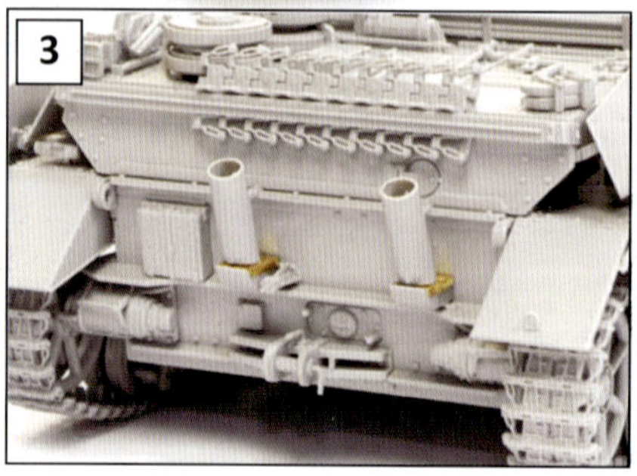

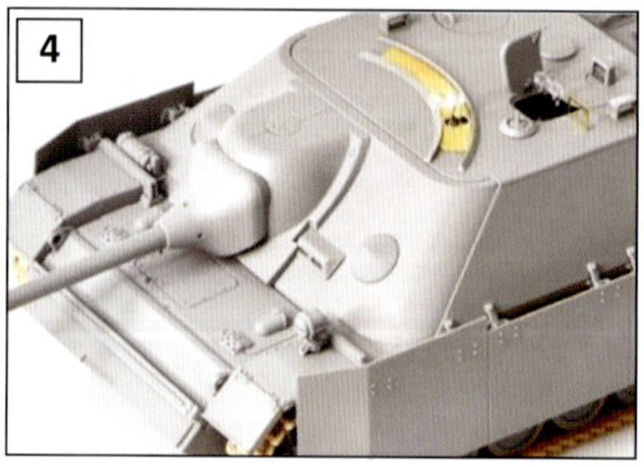

TRUMPETER/HOBBY BOSS

Often thought to be separate entities, these Chinese companies are in fact owned by the same corporation, are located at the same address and are essentially the same firm although different models are marketed separately under the two labels. Hobby Boss produces a 1/35 scale Panzer IV/70(A) which contains a photo-etched brass Thoma Schild which needs to be cut into three separate segments. Trumpeter's range includes a 1/72 scale model of a Jagdpanzer IV which can also be built as a Panzer IV/70(V) tank destroyer. Although this kit would be more than adequate as a wargames model, it is somewhat rudimentary when compared to the offerings from Dragon Models.

Above: The very dramatic box art for the Hobby Boss 1/35 scale Panzer IV/70(A) model executed by the very talented Arkadiusz Wróbel. At right: Details of the kit parts. Note that the Thoma Schild screen is supplied in one piece as mentioned above.

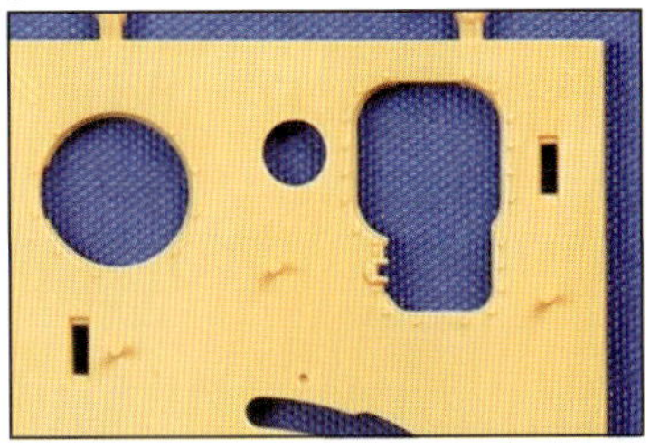

ITALERI

This Italian company has been producing plastic models since its foundation in 1962 and the current catalogue contains a 1/35 scale Jagdpanzer IV ausf F which was released in 2011 and utilised some parts from the company's 1994 Panzer IV/70(V) kit. The latter seems to have been dropped from Italeri's catalogue but is still listed as available by a number of retailers. In 1/72 scale Italeri offers a Jagdpanzer IV ausf F, which features the same box art as the larger scale kit, but is in fact the ESCI model first released in the 1980s with a number of additional parts and new markings. A number of Italeri's armour models have also been released under the World of Tanks label.

Below: Italeri's 1/35 scale Panzer IV/70(V) finished with the photo-etched set from Eduard Model Accessories. This model appears to have been dropped from the company's catalogue. At right: Details of the Italeri Jagdpanzer IV L/48 in 1/35 scale. The photo-etched brass parts and the aluminium barrel are included in the kit.

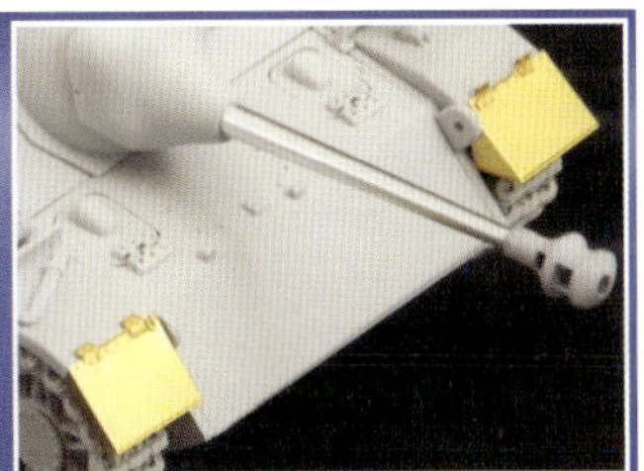

GUNZE SANGYO

This Japanese company was one of the first to release so-called multi-media kits that featured photo-etched brass, resin and aluminium parts in addition to injection-moulded plastic. Controlled today by GSI Creos Corporation, the Gunze company concentrates mainly on model paint, putty and related products. Originally released in 1997, the company's Jagdpanzer IV L/48 kit in 1/35 scale was in fact a re-boxing of Dragon Models' Jadgpanzer IV L/48 and did not contain the additional details for which Gunze Sangyo was known. This kit also formed the basis for Revell's Jagdpanzer IV ausf F Early 1/35 scale model and although apparently out of production is available today, updated with new parts, in a kit which can also be built as the Panzer IV/70(V) tank destroyer.

At left: Vince Wai's artwork for the Gunze Sangyo Jagdpanzer IV. Below: The 1/35 scale kit built and painted by Kentaro Takabayashi. This model is also featured in the Model Showcase section of this book from page 34.

MATCHBOX, REVELL, HASEGAWA

First released in 1978, Matchbox's Panzer Jaeger IV/70 model in 1/76 scale contained fifty-five pieces in injection-moulded plastic and two lengths of track in soft vinyl. As with all the models in the company's Purple Range, a large scenic base on which the vehicle could be displayed is included. An example of the finished model is shown on page 41 of the Model Showcase section of this book. The kit was re-boxed and released by Revell in 2008 as Deutsche Jagdpanzer IV L/70 and in a set containing a number of soft plastic figures and can still be obtained from some retailers although it is not shown in the company's current catalogue. Japanese company Hasegawa's 1/72 scale Jagdpanzer IV L/48 was released as an early version in 2001 and was very quickly followed by a Panzer IV/70 model, largely based on the earlier release. Although they contain some minor inaccuracies, both kits are well detailed and easy to build. The soft plastic tracks, which were almost useless, were apparently replaced by more detailed versions in the 2003 re-release.

Above, from left to right: The original box art for the Matchbox 1/76 scale Panzer Jaeger IV L/70. Revell's re-release of the Matchbox kit under a slightly altered name. Hasegawa's 1/72 scale Jagdpanzer IV early version built and painted straight from the box. Note the extemely rudimentary tracks.

E.T. MODEL

E.T. Model, a relatively new manufacturer based in Shanghai, produce highly-detailed upgrade sets in 1/72 and 1/35 scale in brass, aluminium and resin. Most sets are designed, according to the company, to fit specific kits but some, for example the stowage bin at left, are referred to as universal. The images below all show details of the 1/35 scale upgrade set for the Dragon Models Panzer IV/70(A).

The company also offers sets for the fenders and track guards of the Panzer IV/70(V) and a basic detail set for the same vehicle which includes a finely detailed Sternantenna D for the FuG8 command radio. As this book was nearing completion the managing director wrote to say that a comprehensive detail set, including a new barrel, would be released in late February 2020.

At right and below: Dragon Models' 1/35 scale Jagdpanzer IV/70(A) kit finished with the E.T. Model's photo-etched brass detail set. Note that some resin parts are also included, notably the periscopes and Scherenfernrohr binoculars. Unfortunately the company's upgrade set for the Panzer IV/70(V) was not released in time to include in this book.

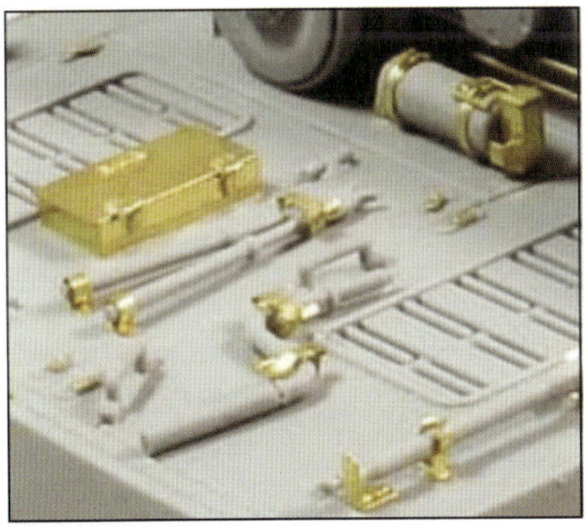

PASSION MODELS

Passion Models is Japanese retailer M.S Models' venture into photo-etched brass detail sets and model accessories. The company also produces high quality waterslide transfers, detailed figures, all in 1/35 scale, and their own glossy magazine. In addition M.S. Models also stock a number of hard to get Japanese products including Kamizukuri and Cobaanii diorama accessories and Twilight Model waterslide transfers. The owner, Mrs Emi Soyama, is an old friend and I can personally recommend their service and products. Shown here are details of Passion Models' photo-etched brass set and turned aluminium barrel for the Panzer IV/70 in 1/35 scale.

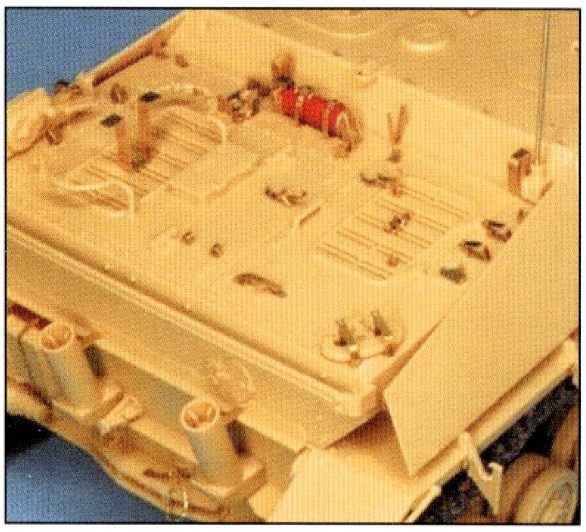

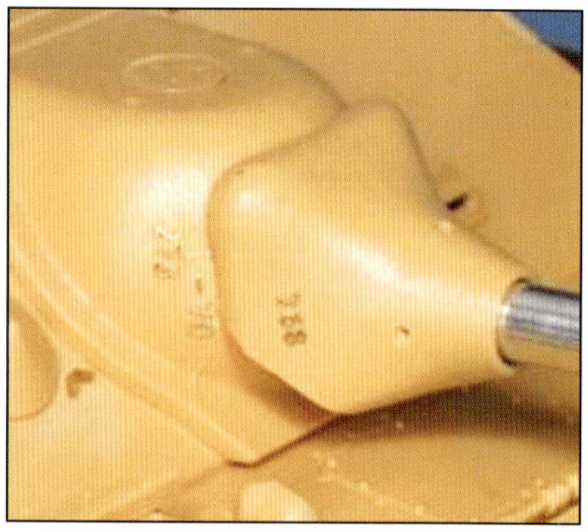

GRIFFON MODEL

Based in China, this company produces a large range of photo-etched brass sets for the Jagdpanzer IV L/48, Panzer IV/70(V) and Panzer IV/70(A) in 1/35 scale. These sets are of a very high quality and usually include a large number of separate pieces. Detailed lists and instructions accompany each set.

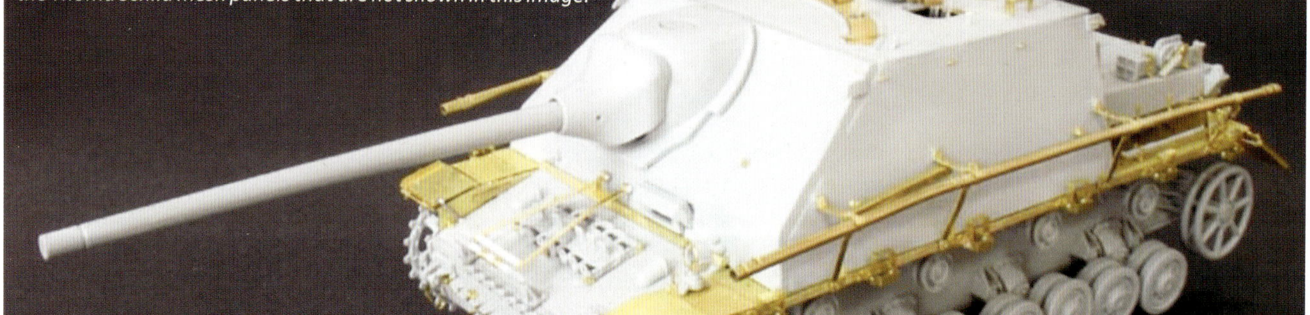

At right: Dragon Models' 1/35 scale Jagdpanzer IV/70(A) kit built and finished with Griffon Model's upgrade set which includes the Thoma Schild mesh panels that are not shown in this image.

Below: Detailed shots of a number of Griffon Models upgrade sets, all in 1/35 scale. 1, 2 and 3. Panzer IV/70(A) showing the complicated bracket arrangement which held the Thoma Schild wire mesh panels and the Flammenvernichter exhaust mufflers. 4. Jagdpanzer IV L/48 with optional command radio insulator and armoured cover. 5 and 6. Panzer IV/70(V) with photo-etched brass and resin parts.

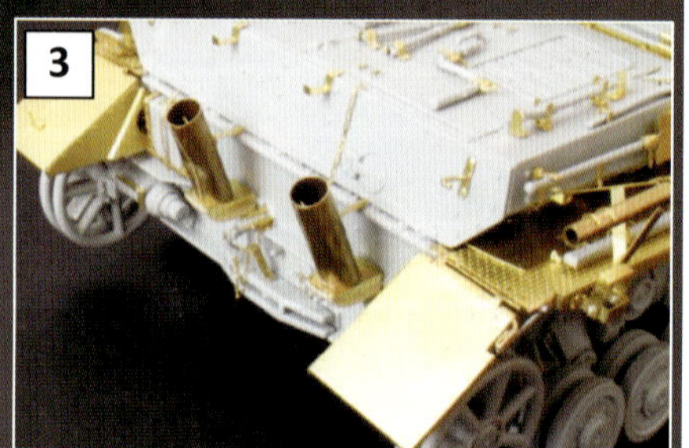

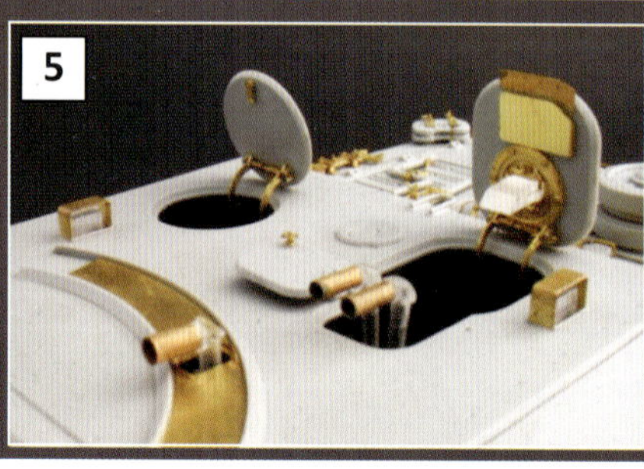

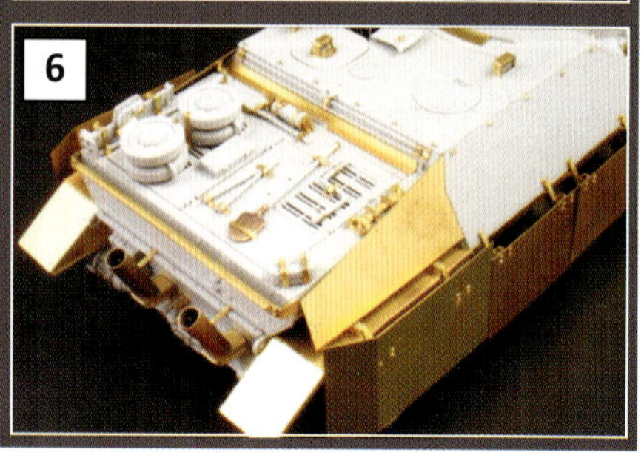

VOYAGER MODEL

Voyager Model, based in Shanghai, have been manufacturing upgrade sets for scale models since 2003 with the release of their first set for 1/35 scale armour. The company produces a number of items in both 1/72 and 1/35 scale in photo-etched brass and also resin including ammunition crates and boxes and fuel containers. Although most of these details would be compatible with almost all manufacturers' models, some sets are marketed as appropriate for specific kits. Shown below, all in 1/35 scale are: 1 and 2. Parts of the upgrade set for Tamiya's Panzer IV/70(V). 3 and 4. Dragon Models' Jagdpanzer IV L/48 model built with Voyager Model's photo-etched brass set. 5 and 6. Details of Voyager Model's upgrade set for the Dragon or Tristar/Hobby Boss Panzer IV/70(A) model.

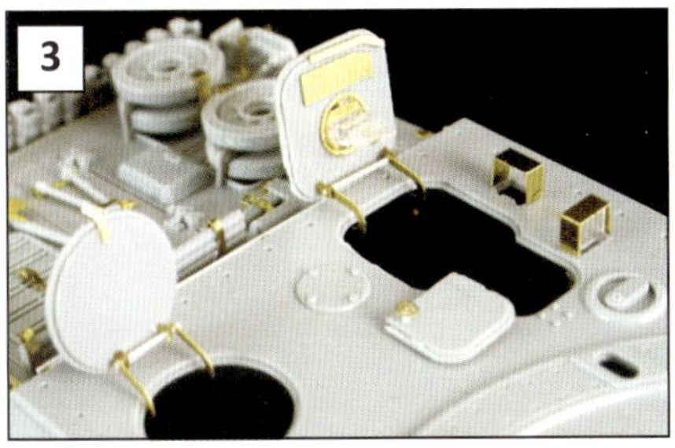

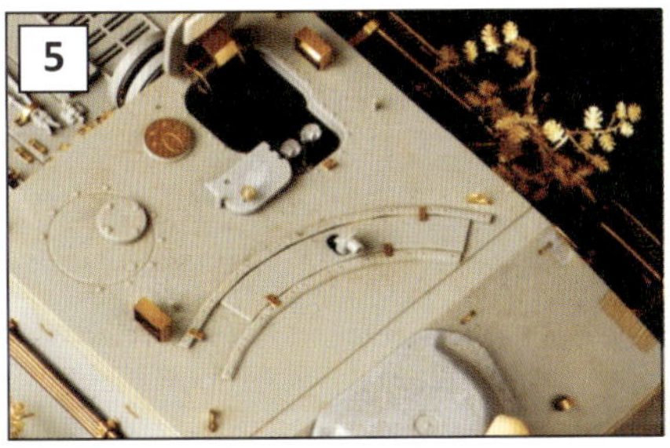

ABER

This Polish company has been manufacturing and selling upgrade sets since 1995, working in photo-etched brass, milled aluminium and brass, stainless steel and even wood. For some time now Tamiya has included Aber products with a number of their models. A full list of the accessories made specifically for the Jagdpanzer IV and Pzkpfw IV in all scales would be far too large to reproduce in this book and the reader can find the company's contact details on page 64. Show here are details of the 1/35 scale upgrade set for Dragon Models' Jagdpanzer IV A-O, the prototype vehicle.

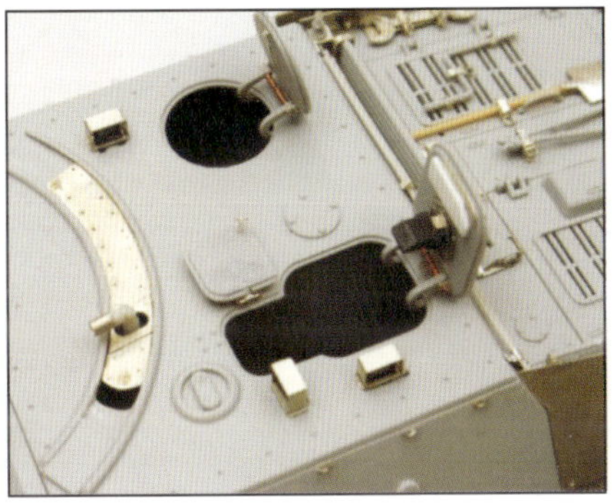

RB MODEL

RB Model from Poland produce a range of milled aluminium and brass gun barrels, ammunition and resin accessories for 1/35, 1/48 and 1/72 scale armour models. Shown here below, from left to right, are shells for the 7.5cm L/48 gun and the 7.5cm L/40 with armour-piercing, high explosive and sabot rounds. Each set contains twelve rounds. The muzzle brake of the 7.5cm L/48 gun. This set contains five parts including the aluminium barrel. All items are in 1/35 scale

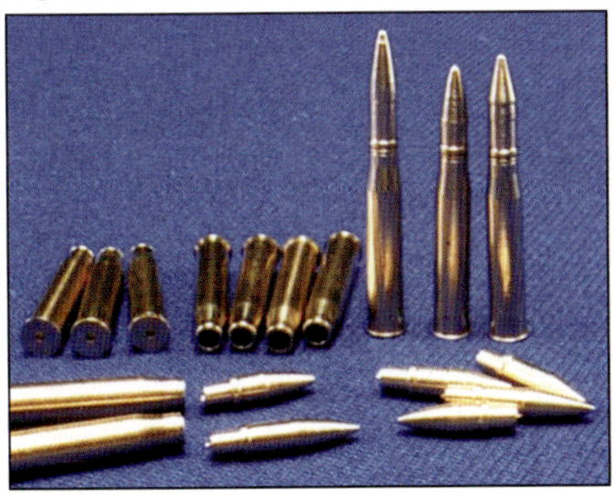
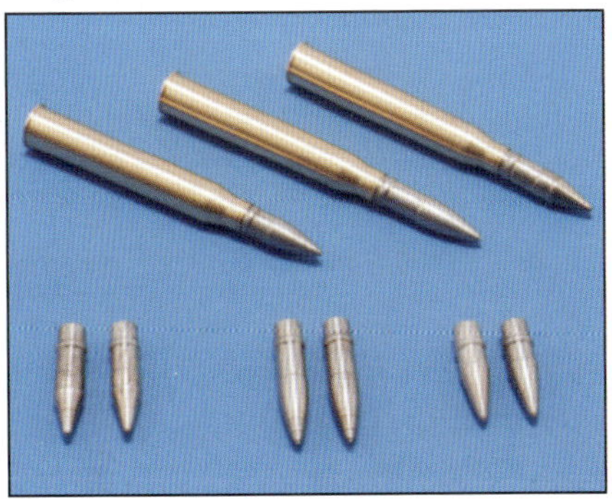

EDUARD MODEL ACCESSORIES

This Polish company has been manufacturing high-quality accessories in photo-etched brass and resin for many years and was one of the earliest firms to offer upgrade sets in the smaller scales. Eduard offers a number of detail sets marketed as appropriate for the Tamiya, Dragon and Italeri kits although they would be compatible with other manufacturers' products. In addition parts of the upgrade sets for the Pzkpfw IV ausf J, particularly the fenders and trackguards would also be suitable.

Below: The Tamiya 1/35 scale Jagdpanzer IV 70(V) model finished with Eduard Model Accessories' photo-etched brass set.

Below: 1. Parts from Eduard's upgrade set for the Tamiya 1/35 scale Jagdpanzer IV 70(V) model. 2, 3 and 4. Italeri's Jagdpanzer IV L/48 kit finished with Eduard Model Accessories' photo-etched brass set. Note the hooks on the hull side which hold the Schürzen.

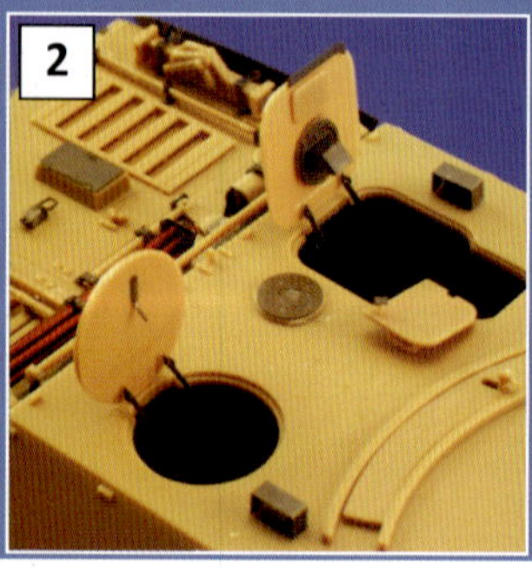

AFTERMARKET TRACKS

Most plastic kits today contain realistic tracks made from a form of vinyl which, unlike the older types, can be glued together, or hard plastic individual links. Nevertheless, there are many modellers who wish for the highest level of detail and realism possible and consequently the number of quality aftermarket products available seems to have increased in recent years, despite the extra expense and work required. Motorised models require workable tracks made up from individual links held together by pins, as were the full size items. The examples shown here, all in 1/35 scale, are 1. Later Pzkpfw IV tracks in white metal from Russian company Master Club. 2. Pzkpfw IV tracks in plastic from ModelKasten. 3. Individual Pzkpfw IV track links in white metal and brass from Friulmodel showing the inside face and the method of linking the tracks.

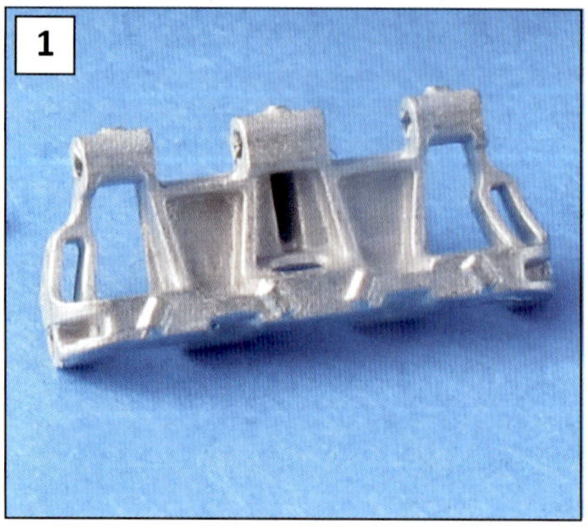

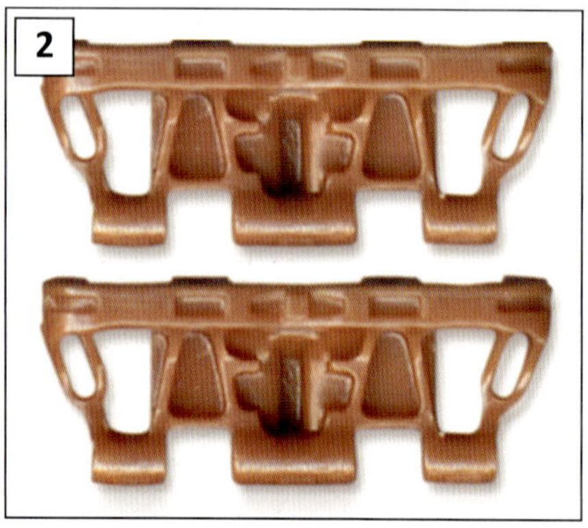

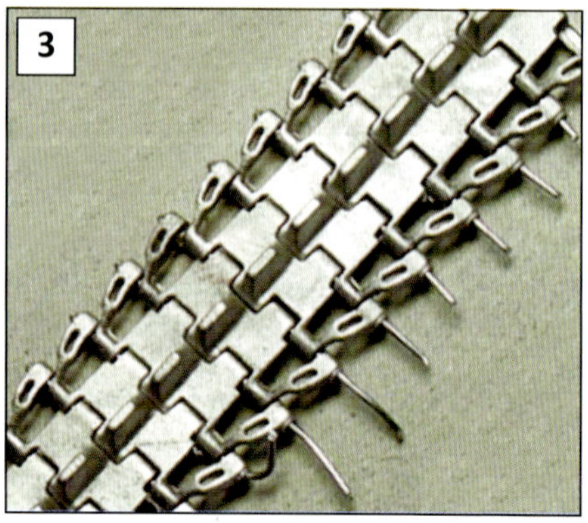

Photographed at Oldenburg in northern Germany in early May 1945, both Panzer IV/70(V) tank destroyers are from 3.Kompanie, schwere Panzerjäger-Abteilung 655. The town was used as a collection point for German vehicles after the surrender and the Marder II in the background probably has no connection to the battalion. These vehicles are very late production models and may be two of the six tank destroyers received in early April 1945.

Late 1943 saw the introduction of the Nashorn self-propelled 8.8cm anti-tank gun and these powerful weapons were allocated to newly created independent formations referred to as schwere Panzerjäger-Abteilungen or heavy anti-tank battalions. A number of these units were later reorganised and rebuilt with Jagdpanther tank destroyers but due to production delays, in September 1944 Hitler ordered that the shortfalls could be made up with Sturmgeschütz assault guns or Jagdpanzer tank destroyers. These units are covered in detail in *Jagdpanther Tank Destroyer: German Army and Waffen-SS, Western Europe 1944-1945,* the eighth book in this series, and those that served on the Western Front are mentioned only briefly here.

Schwere Panzerjäger-Abteilung 93. Equipped with Nashorn self-propelled anti-tank guns, this battalion was almost completely destroyed on the Eastern Front in August 1944. The plan to rebuild the battalion as a Jagdpanther unit was never realised and at the end of November the first company was hurriedly equipped with Nashorns and attached to a Kampfgruppe controlled by the headquarters of Panzer-Regiment 2 (see above).

On 17 February 1945 the battalion's 2.Kompanie was ordered to join Panzer-Brigade 106 and was allocated the eight Panzer IV/70(A) which had been left at the front by 7.Kompanie, Panzer-Regiment 2 (1). On 15 March 1945 the brigade headquarters reported that four of these vehicles were still operational.

Schwere Panzerjäger-Abteilung 519. The battalion served on the Russian Front until July 1944 when it was withdrawn to Truppenübungsplatz Mielau in Germany to be rebuilt. Renamed schwere Panzerjäger-Abteilung 519 (Panther) on 15 September 1944, the battalion was organised with a Stabskompanie of three Jagdpanthers, 1.Kompanie with fourteen Jagdpanthers and the second and third companies with fourteen Sturmgeschütz III assault guns each. The Jagdpanthers were sent into action almost immediately, supporting I.SS-Panzerkorps and 116.Panzer-Division in the battles for the Hürtgen Forest.

In early December the battalion was attached to 246.Volksgrenadier-Division, which had been fighting around Aachen, and by this time nine Jagdpanzer IV/70(V) tank destroyers had arrived to supplement the assault guns, of which nineteen remained. In March the battalion was subordinated to 1.Armee, which was retreating from the area around Kaiserslautern towards the Danube, and by early April all the armoured vehicles had been lost

Schwere Panzerjäger-Abteilung 559. Formed on 10 April 1944 at Truppenübungsplatz Mielau and equipped with Jagdpanther tank destroyers, the battalion was continuously engaged on the Western Front from September 1944. From 11 December 1944, the battalion was attached to Panzer-Lehr-Division and by that time the first company was equipped with fourteen Jagdpanthers while 3.Kompanie had been issued the battalion's first fifteen Jagdpanzer IV/70(V) tank destroyers earlier that month. A further three Jagdpanzers almost certainly went to the battalion headquarters while 2.Kompanie controlled the remaining fourteen Sturmgeschütz III assault guns. In March 1945 the battalion was subordinated to XIII. SS Armeekorps fighting to defend the Saarpfalz region.

Schwere Panzerjäger-Abteilung 560. The battalion had served in Russia during 1943, equipped with Nashorn 8.8cm self-propelled guns, until May 1944 when

Notes

1. I have assumed this total based on the number of tank destroyers issued to Panzer-Regiment 2 and the subsequent losses. This unit is sometimes referred to as schwere Heeres-Panzerjäger-Abteilung 93

its personnel were transferred to Germany to begin conversion to the Jagdpanther. The first allocation of vehicles did not reach the battalion until October 1944 by which time Hitler had decided that the heavy Panzerjäger battalions in training should be equipped with a single company of Jagdpanthers and two companies of Sturmgeschütz III assault guns or Jagdpanzer tank destroyers. The first shipment of eleven Jagdpanzer IV L/70 (V) tank destroyers was despatched in early October while a further fifteen were sent in November and a further two in December. During the same period the battalion was also allocated a total of eleven Jagdpanthers.

The battalion took part in the Ardennes Offensive as part of Kampfgruppe Kuhlmann which was largely made up of units from 12.SS-Panzer-Division and although schwere Panzerjäger-Abteilung 560 was probably under strength at this time, figures given in some accounts of a single Jagdpanther and four Jagdpanzer IV as combat ready are almost certainly incorrect. The battalion was engaged in the assault on Oubourcy, north-east of Bastogne, and in the fighting for control of the Büllingen to Bütgenbach road. In January 1945, the battalion was fighting in the Rhine Brückenkopf area near Gambsheim and Hagenau in support of 10.SS-Panzer-Division and in the following month was transferred to the Eastern Front where it remained until the end of the war fighting in Hungary and on the Oder Front.

Schwere Panzerjäger-Abteilung 655. The battalion had served on the Russian Front until 13 September 1944 when it was reorganized at Truppenübungsplatz Mielau as a schwere Panzerjäger-Abteilung (Panther). Although some sources state that the third company remained in Russia, I have not been able to confirm this and it is certain that the battalion was made up of three full companies at Mielau.

The first five Jagdpanthers were shipped from the Heereszeugamt on 24 November 1944 and these were allocated to 2.Kompanie. On the following day twenty-eight Jagdpanzer IV/70(V) tank destroyers were despatched for the first and third companies and in early December a further three were allocated to the battalion headquarters. During the first week of December 1944, the first and third companies left for the front while the second company remained at Mielau, waiting for their last delivery of Jagdpanthers which were not shipped until Christmas Eve. In early 1945 the battalion's 3.Kompanie was detached, re-equipped with Hornisse self-propelled anti-tank guns and renamed schwere Panzerjäger-Kompanie 669. The company was sent to the east and took part in the battles in Czechoslovakia. From the beginning of March 1945 the remainder of the battalion was attached to Korps-Bayerlein, part of 15.Armee holding the front north of Koblenz (1). The battalion received six new Panzer IV/70(V) tank destroyers in early April and these must have been among the last of these vehicles allocated to combat units, indeed they may have originally been intended for another formation. The battalion surrendered to the Canadians in on 7 May 1945 near Oldenburg, north-west of Bremen.

Notes

1. Also referred to in some accounts as Division-Bayerlein, this formation was built around the remnants of LIII.Armeekorps.

Although any positive identification of this Panzer IV/70(V) tank destroyer is not possible, the style and placement of the company number is very similar to the tank destroyers of 3.Kompanie, schwere Panzerjäger-Abteilung 655 depicted on the previous page. Note the Absatzkrümmer curved extensions fitted to the Flammentöter exhaust mufflers and the starting crank still in place on the hull rear. This vehicle is also shown on page 21 of the Camouflage & Markings section of this book.

SCHWERE PANZERJÄGER-ABTEILUNG, SEPTEMBER 1944

The initial organisation of a schwere Panzerjäger-Abteilung (Panther) called for a Stab, or headquarters, and a Stabskompanie which controlled a signals platoon, an armoured engineer platoon, an anti-aircraft platoon and a medical platoon. It is unlikely that all battalions had their full complement of these support units. The battalion was to contain three fighting companies, each equipped with fourteen Jagdpanthers. In early September, confirming a practice that had apparently been in place since late August, a new organisation was introduced where only the first company was to be equipped with Jagdpanthers while the second and third companies were outfitted with Sturmgeschütz III assault guns or Jagdpanzer tank destroyers. Details of these allocations are given in the text. The expression 'frei Gliederung', sometimes rendered in official documents as simply 'frei', refers to the concept of free organisation which was introduced in early 1944 where certain support elements, such as transport or supply, were removed from companies and held at battalion or regimental level. Two battalion command vehicles were equipped with Funkgerät (FuG) 8 and FuG5 radios while the third Jagdpanther had FuG7 and FuG5 sets. Each company commander's vehicle was equipped with an FuG8 and FuG5 radio and the four platoon commanders had FuG5 and FuG2 sets. The remaining vehicles were fitted with a single FuG5 radio. The FuG8 radio is easily identified by the Sternantenna aerial.

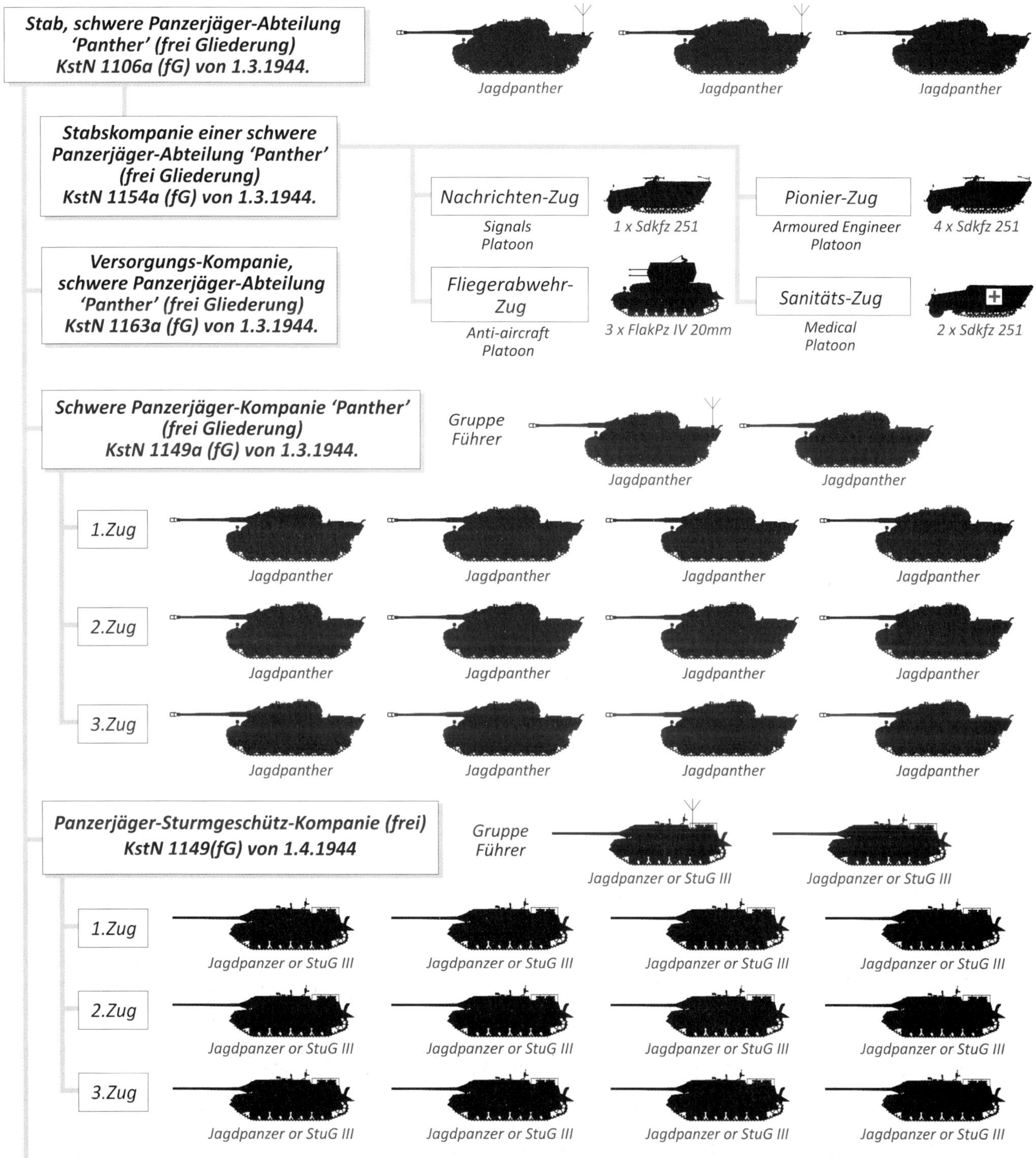

Panzerjäger-Sturmgeschütz-Kompanie (frei)
KstN 1149(fG) von 1.4.1944

All the schwere Panzerjäger units that served on the Western Front were allocated either Sturmgeschütz III assault guns or Panzer IV/70(V) tank destroyers as described in the main text. It seems that, where possible, the Ausfuhrung B version of KstN 1149(fG), which allowed for fourteen-vehicle companies, was used.

As mentioned earlier, the Jagdpanzer series was developed to eventually replace the army's Pzkpfw IV medium tank and to strengthen the Panzerjäger battalions. Although these vehicles were essentially assault guns it was never envisaged that they would be allocated to Sturmgeschütz units and this had as much to do with the rivalry between the artillery inspectorate, which controlled the Sturmgeschütz battalions, and the Generalinspekteur der Panzertruppen as it did with any logistical practicalities. Nevertheless, during the first three months of 1945, seventeen Sturmgeschütz battalions, serving on the Eastern and Western Fronts, were each promised a platoon of Panzer IV/70(A) tank destroyers. Many of these never arrived at their destinations and some were appropriated by other units. In addition a number were also shipped to Sturmgeschütz training and replacement units (1). The battalions that fought in the West are listed below.

The assault gun units that fought in the West during the last year of the war are examined in greater detail in *Stug III and Stug IV: German Army, Waffen-SS and Luftwaffe, Western Front, 1944-1945,* the nineteenth book in the TankCraft series.

Sturmgeschütz-Brigade 243. After losing all its vehicles in January 1945, the battalion was rebuilt in March and April and received the last Sturmgeschütz III assault guns manufactured by Alkett at the firm's Berlin-Borsigwalde plant. At that time the battalion was renamed Heeres-Sturm-Artillerie-Brigade 243 and was supposed to have received a platoon of three Panzer IV/70(A) tank destroyers but I have been unable to confirm if these were delivered. On 15 March 1945 the battalion reported that none were on hand.

The battalion was subordinated to 12.Armee and in mid-April 1945 took part in the attacks on the American bridgehead at Schönebeck, south-east of Magdeburg.

Sturmgeschütz-Brigade 244. In October 1944 the battalion was rebuilt at Truppenübungsplatz Altengrabow, near Magdeburg in Germany, after being completely wiped out during 1944 in the battles around Bobriusk, south-east of Minsk in modern-day Belarus. In December the battalion took part in the Ardennes Offensive with slightly more than a single battery of guns while just seven were listed as serviceable at the end of the month.

The battalion was allocated three Panzer IV/70(A) in January 1945 but it is possible that they were never received and they may have been allocated to other units (2). The battalion was engaged in the defence of Cologne and ended the war fighting in the Ruhr Pocket.

Sturmgeschütz-Brigade 280. Formed in August 1943 as Sturmgeschütz-Abteilung 280, the battalion served in the East and was renamed Sturmgeschütz-Brigade 280 on 14 February 1944. Withdrawn to Denmark after the fighting around Tarnopol, now Ternopil in modern-day Ukraine, the battalion was completely rebuilt.

The battalion's 3.Batterie, commanded by Major Kurt Kühme, famously took part in the battle for Arnhem in September 1944. The battalion was later involved in the fighting in northern Liechtenstein and the battles for the Moerdijk Bridgehead and by the end of November had been transferred to Alsace where it remained until February 1945.

The battalion was allocated three Panzer IV/70(A) tank destroyers on 28 January 1945. In March 1945 the battalion reported that six Panzer IV/70(A) were on hand, four of which were fully operational, and I can offer no explanation as to where the additional three vehicles originated. After taking part in the counterattacks in Alsace, the battalion was involved in the defensive battles between Speyer and Wasserburg am Inn. On 26 April 1945 the battalion had reached Bopfingen, between Stuttgart and Nuremburg, and although the men were ordered to make their own way home, most were taken into captivity.

Sturmgeschütz-Brigade 341. Formed in May 1943, the battalion was stationed in southern France when the Normandy landings began and suffered heavily from its first actions. By 20 August, despite the arrival of some replacements, the battalion could report that only twenty vehicles were on hand and within five days this number was reduced to twelve. In September the battalion was withdrawn to Aachen and rebuilt and took part in the fighting in the Hürtgenwald.

In January 1945 three Panzer IV/70(A) tank destroyers were shipped from the Heereszeugamt and these arrived in the following month. Curiously, the battalion consistently reported that four Panzer IV/70(A) were on hand as late as 15 March 1945. and these may have been the unit's last serviceable vehicles.

The battalion remained in the Sauerland region into the early months of 1945 but managed to escape the Ruhr Pocket, retreating towards Kassel and the area around Magdeburg where most of the battalion surrendered to the Americans on 20 April 1945.

Sturmgeschütz-Brigade 394. Formed in May 1944 as Sturmgeschütz-Abteilung 394 for the proposed 18.Artillerie-Division, the battalion was renamed on 10 June 1944. The battalion took part in the Normandy battles, most notably at Vire in early August, and was later trapped in the

Notes

1. These were eventually sent to the front but as they served in the East they need not concern us here.
2. Interestingly, the battalion reported that two Flakpanzers were on hand in early 1945.

Falaise Pocket. The remnants were then withdrawn to Aachen and the battalion was rebuilt. The battalion took part in the Ardennes Offensive and was engaged in the battles in Alsace and Lorraine in January 1945. In the same month three Panzer IV/70(A) were shipped to the battalion by the Heereszeugamt but on 26 January, the date of their expected arrival, Sturmgeschütz-Brigade 394 was rushed to Saarbrucken where it was subordinated the Heeresgruppe G. In addition, they are not mentioned in the 15 March 1945 report and it is unlikely that the vehicles ever arrived at their original destination.

By April the battalion had lost all its assault guns and the surplus crews were incorporated into Infanterie-Division Ferdinand von Schill.

Sturmgeschütz-Brigade 667. Originally formed as Sturmgeschütz-Abteilung 667 in June 1942, the battalion was renamed Sturmgeschütz-Brigade 667 in February 1944 and finally Heeres-Sturmartillerie-Brigade 667 sometime in the following June. The battalion was almost completely destroyed in the summer 1944 battles on the Eastern Front and was withdrawn to Holland and rebuilt. In September 1944 the battalion was attached to Kampfgruppe Chill, an ad-hoc formation scraped together from elements of 85.Infanterie-Division, and may have had as many as nineteen Sturmgeschütz III assault guns and twelve Sturmhaubitze 42 on hand.

The battalion took part in the Ardennes Offensive in December 1944 as part of 6.Panzerarmee, reporting that just five assault guns were operational. In February 1945 the battalion was allocated three Panzer IV/70(A) tank destroyers and these arrived in the following month. In a report prepared for the Generalinspekteur der Panzertruppen dated 15 March 1945 the battalion reported that six Panzer IV/70(A) were on hand and all were fully operational, suggesting that they were recent arrivals. On 18 March, in what may have been one of the battalion's last actions, fifteen assault guns managed to hold up an American attack at Bad Kreuznach until they were driven back by Allied fighter-bombers. Following this battle the battalion retreated towards Volxheim and Worms on the Rhine, abandoning a number of vehicles along the way until just three assault guns remained (1).

Notes

1. The available official records do not describe the battalion's history during the last months of the war and this passage is based on the account of Herr Anton Schupp who served as an Obergefreiter with the battalion.

Photographed in Bennerscheid, a small village south-east of Bonn, in late March 1945 this Panzer IV/70(A) tank destroyer may be one of the vehicles operated by Sturmgeschütz-Brigade 341. Part of the sight of the Kugellafette Vorsatz P mount can be seen on the open loader's hatch. Note the rack on the superstructure side and the rail around the engine deck, presumably for stowage. These may have been modifications that were introduced very late in the production run. This image also provides a good view of the Drahtgeflecht wire mesh panels, sometimes referred to as Thoma Schild, and the system of rails and brackets used to hold them in place. Note that the Schürzen on the left-hand side seems to consist of metal plates. This vehicle is also shown and discussed on page 26 of the Camouflage & Markings section.

Photographed outside Volxheim, south-east of Bad Kreuznach, this Panzer IV/70(A) may have one of the vehicles of Heeres-Sturmartillerie-Brigade 667 or Heeres-Sturmgeschütz-Brigade 280 which were both operating in this area in March 1945. Although of mediocre quality, this image, and others in the series, show features which may have been standard on very late production vehicles but not mentioned in official documents. These include a rack to hold lengths of spare track on the right-hand side of the superstructure and a stowage rail on the rear deck, similar to that fitted to the Panzer IV/70(A) on page 53 but not visible here. It is also equipped with the Vorsatz P mount on the loader's hatch and an armoured cover on the rear of the fighting compartment for the FuG8 command radio.

Notes

1. The development of the Tiger and the competing designs are examined in detail in *Tiger I and Tiger II: Tanks of the German Army and Waffen-SS, Eastern Front 1944.*

2.These features are explained further in the Technical Details & Modifications section of this book.

3. This was in fact the first of the battalion's tanks that was lost on the Western Front. The first personnel loss was incurred on 28 June when Unteroffizier Otto Oelsner of 1.Kompanie was killed in an accident as the battalion was transported by rail to the front.

Sturmgeschütz-Brigade 902. Formed in January 1944 as the Lehr-Abteilung of the Sturmgeschütz-Schule West at Tours, and renamed on 14 February 1944, the battalion took part in the Normandy campaign and the fighting around Aachen.

In early December 1944, Sturmgeschütz-Brigade 902 was attached to Korps Feldt, an ad-hoc unit of 1.Fallschirm-Armee, and later that month took part in the Ardennes Offensive as part of 6.Panzerarmee. In January 1945, three Panzer IV/70(A) tank destroyers were shipped from the Heereszeugamt and although these arrived in the following month the battalion reported on 15 March that four vehicles were on hand of which three were operational. Although some sources suggest that the battalion was transferred to the East in mid-January 1945 surviving prisoner interrogation statements compiled by the US Army indicate that in early April the second and third batteries were supporting 176.Volksgrenadier Division near Brilon on the eastern edge of the Ruhr Pocket while the first battery was almost 50 kilometres to the south-west at Detmold with five serviceable vehicles. All the batteries were caught in the Ruhr Pocket and by 8 April 1945 the last Sturmgeschütz had been lost.

Sturmgeschütz-Brigade 905. The battalion was badly depleted by the battles on the Eastern Front in June 1944 and withdrawn to Germany to be completely rebuilt. Renamed Heeres-Sturmartillerie-Brigade 905 in the autumn of 1944, the battalion was attached to 5.Panzerarmee in the Eifel in early 1945. At that time Heeres-Sturmartillerie-Brigade 905 had twenty-two Sturmgeschütz III assault guns and nine Sturmhaubitze 42 on hand. The battalion was allocated three Panzer IV/70(A) in January 1945 but it is possible that they were never received. It seems likely that the vehicles which could not be delivered to units such as Sturmgeschütz-Brigaden 244, 394 an 905 were used to reinforce the battalions that could be reached by the overtaxed transport system. In March the battalion reported that no assault guns of any type were on hand and an OKH order from the same month states that it would not be possible to return the battalion to Germany to be refitted.

Sturmgeschütz-Brigade 911. The battalion was rebuilt in August 1944 with three assault gun batteries after being destroyed in the fighting in Romania. In December the battalion was renamed Heeres-Sturmartillerie-Brigade 911 and temporarily attached to the Führer-Grenadier-Brigade in anticipation of the Ardennes Offensive. The battalion was allocated three Panzer IV/70(A) tank destroyers in late January 1945 and some sources state that these arrived at some time in February. However, they are not mentioned in any detailed reports compiled during the battalion's later service in the East and it is entirely possible that they never arrived.

The battalion was officially absorbed into the Führer-Grenadier-Division on 2 March 1945 and renamed Führer-Sturmartillerie-Brigade.

The notion that the units of the Waffen-SS received preferential treatment in the allocation of men and material is something of a myth and of the six divisions which fought in the Normandy battles, all were understrength from the beginning of the campaign and by the time of the Mortain counterattack in August 1944 had been reduced to weak Kampfgruppen.

All were partially rebuilt in October and November but suffered heavy losses in equipment during the Ardennes Offensive and Operation Nordwind. Those formations which were issued Jagdpanzer tank destroyers are listed here.

SS-Panzerjäger-Abteilung 1. Formed in October 1943, the battalion was attached to 1.SS-Panzer-Division Leibstandarte SS Adolf Hitler (1). In early 1944 the battalion was used to form SS-Panzerjäger-Abteilung 12 which was subsequently destroyed in the Normandy battles.

Rebuilt and attached to its former division in late 1944, the battalion was allocated twenty-one Panzer IV/70(V) tank destroyers on 20 October, although these probably arrived in the following month. The battalion took part in the Ardennes Offensive as part of Kampfgruppe Hansen and it is likely that eighteen operational vehicles were available at the commencement of the attack (2). I have been unable to locate any exact figures detailing the battalion's losses during this operation but a report dated 15 January 1945 shows that the division had eleven Jagdpanzers and four assault guns on hand with none in repair (3). On 6 January 1945 the battalion was assigned two new Panzer IV/70(V) tank destroyers and during the first week of February the division was transferred to the Hungary and took part in Operation Spring Awakening, the last major German offensive of the war.

SS-Panzerjäger-Abteilung 2. Formed on Friday, 8 September 1944 from SS-Sturmgeschütz-Abteilung 2 the battalion was attached to 2.SS-Panzer-Division Das Reich. Although it was planned that the battalion would be made up of a headquarters with three Jagdpanzer IV and two companies each equipped with fourteen similar vehicles, these figures were never attained and twenty Panzer IV/70(V) tank destroyers were allocated in November 1944 (4) while a further three were allocated in December.

On 15 December, on the eve of the Ardennes Offensive, the battalion reported that just eight of these vehicles were considered fully operational while the remainder were in short-term repair. The battalion took part in Operation Wacht am Rhein, the assault through the Ardennes, as part of II.SS-Panzerkorps with 9.SS-Panzer-Division.

On 17 January 1945, after fighting in the battles between Magoster and Marcourt in central Belgium, the division was withdrawn from the front and moved behind the Westwall defensive line to Koblenz to rest and refit. In the following

Notes

1. As a number of Waffen-SS formations went through several name changes during the war I have used the titles that were current in the summer of 1944, when our study begins.
2. Some sources suggest that all twenty-one tank destroyers took part in the initial advance and others as few as ten. I feel the serviceable figure of eighteen given here is probably correct while the number of ten may refer to 1.Kompanie.
3. At least one vehicle was lost when the Jagdpanzer of Obersturmführer Otto Holst, the commander of 1.Kompanie, crashed through a bridge into the Amblève River near Trois-Points on 21 December 1944.
4. Some sources suggest that the actual figure was as high as twenty-eight and this would have been enough to equip the two companies while the December allocation would have brought the battalion to full strength.

The Panzer IV/70(V) tank destroyer of 1.Kompanie, SS-Panzerjäger-Abteilung 1 commanded by Obersturmführer Holst photographed after it crashed through the bridge over the Amblève River. The so-called Ambush scheme camouflage pattern which was applied to many, but not all, the battalion's vehicles can be seen here. Note the lack of any marking other than the Balkenkreuz national insignia.

month 2.SS-Panzer-Division was transferred to the East where it remained, the last survivors surrendering at Passau. This unit should not be confused with SS-Panzerjäger-Ausbildungs-Abteilung 2 which took part in the fighting in Holland in September 1944.

SS-Panzerjäger-Abteilung 9. Attached to 9.SS-Panzer-Division Hohenstaufen, the battalion was allocated twenty-one Jagdpanzer IV tank destroyers but as the first were not shipped until 29 July 1944 they did not take part in the Normandy battles. Instead, an ad-hoc 3.Kompanie was formed by drafting men from other elements of the division and equipped with towed 7.5cm Pak 40 anti-tank guns. By the end of August all the division's tanks and assault guns had either been destroyed or were handed over to Panzer-Regiment 15 of 11.Panzer-Division.

The Panzerjäger battalion had completed its training and began moving towards the front, arriving at Mons in Belgium on 2 September 1944, but was immediately caught up in a series of airstrikes losing a number of vehicles and the commander of 1.Kompanie, Hauptsturmführer Günter von Brocke.

Reduced to the strength of a weak Kampfgruppe, the division famously took part in the defence of Arnhem in mid-September 1944 but due to the losses of the preceding weeks and the number of Jagdpanzers present may have been as few as two. On 10 October 1944 the battalion reported that just one operational Jagdpanzer IV was on hand. Rebuilt in November with twenty-one Panzer IV/70(V) tank destroyers, the battalion took part in the Ardennes Offensive as part of 6.Panzerarmee and at the end of January 1945, with the remainder of 9.SS-Panzer-Division, was withdrawn from the front to rest and refit. In February, 9.SS-Panzer-Division was sent to the East and was engaged in Operation Spring Awakening, the last major German offensive of the war, and received twelve Panzer IV/70(V) tank destroyers. This represented the battalion's last allocation of the war.

On 1 May 1945 a large part of the division, which had managed to make its way to the West, surrendered to the Americans.

SS-Panzerjäger-Abteilung 10. Attached to 10.SS-Panzer-Division Frundsberg, the battalion was created in June 1944 at the Mielau training facility in Poland but as the first Jagdpanzer IV tank destroyers did not arrive until late August and the battalion was not able to rejoin its parent formation until September.

The division had been badly mauled in the Normandy battles and was withdrawn to the area around Arnhem in Holland and re-organised as Kampfgruppe Frundsberg when the Allies launched Operation Market-Garden. The Panzerjäger battalion, however, was almost at full strength with the first and second companies equipped with ten Jagdpanzer IV each with one tank destroyer assigned to the battalion headquarters (1). After the fighting for Arnhem the division remained in central Holland until November when it was moved to the Ruhr taking part in the unsuccessful attempt to retake the town of Geilenkirchen, north of Aachen.

In the first half of December 1944 the battalion received ten new Panzer IV/70(V) tank destroyers and reported that three Jagdpanzer IV were still on hand, although this may represent the combat-ready figure. In January 1945, 10.SS-Panzer-Division took part in the fighting in Alsace as part of Operation Nordwind, the last major German offensive in the West but by the first week of February the division had been withdrawn from the front and transferred to the East.

SS-Panzerjäger-Abteilung 12. Formed from SS-Panzerjäger-Abteilung 1 in February 1944 the battalion was attached to 12.SS-Panzer-Division Hitler Jugend. Ten Jagdpanzer IV were allocated to the battalion on 26 April 1944 but these took almost a full month to arrive at their destination. These vehicles were split evenly between the first and second companies to facilitate the training of their crews. The battalion reported that the muzzle brakes of all the Jagdpanzers were missing and they all suffered numerous mechanical breakdowns.

By 6 July 1944 all ten vehicles had been concentrated in 1.Kompanie which was very soon after sent into action. Earlier, on 22 June another eleven Jagdpanzer IV were allocated to the battalion and these were picked up in Versailles on 13 July 1944. Ten vehicles were assigned to 2.Kompanie and one Jagdpanzer went to the battalion headquarters. The third company was equipped with 7.5cm Pak 40 towed anti-tank guns.

The division was almost completely destroyed in the Normandy fighting and rebuilt from October 1944 with the Panzerjäger battalion receiving twenty-one Panzer IV/70(V) tank destroyers. The division took part in Ardennes Offensive in December 1944 and was involved, with disastrous results, in the frontal assault on the town of Bastogne, losing seventeen Jagdpanzers and thirteen tanks in this action alone.

The division was withdrawn from the front on 10 January 1945 and after a brief period of rest was sent to Hungary in the first week of February. At that time, according to Hubert Meyer who temporarily commanded the division, the Panzerjäger battalion could field a total of

Notes

1. The towed anti-tank guns, which would have equipped 3.Kompanie, would not be available for some time.

This Jagdpanzer IV L/48 of 2.Kompanie, SS-Panzerjäger-Abteilung 17 was disabled during the fighting near Le Mans in August 1944. Note the early rear idler, cylindrical exhaust muffler and Zimmerit coating. Another vehicle of this company is shown on page 17 of the Camouflage & Markings section of this book.

thirteen tank destroyers, six with 1.Kompanie and seven with 2.Kompanie, while the towed guns were operated by 3.Kompanie. He makes no mention of the headquarters. If these figures are correct the vehicles may have been the survivors of the fighting in the Ardennes.

SS-Panzerjäger-Abteilung 17. Formed in early June 1944 by renaming SS-Sturmgeschütz-Abteilung 17, the battalion was attached to 17.SS-Panzergrenadier-Division Götz von Berlichingen.

When the Allied landings began in Normandy the battalion's 3.Kompanie was equipped with twelve Marder III self-propelled guns, both the 7.62cm and 7.5cm armed versions, while the other companies and the battalion headquarters were without any vehicles. The division was only able to form a weak Kampfgruppe that took part in the early fighting while most of the Panzerjäger battalion remained at Thouars, south of Angers in western France, until the end of June when thirty-one Jagdpanzer IV tank destroyers arrived.

The full battalion went into action for the first time on 5 August 1944 in an attack towards Laval on the Rennes to Le Mans road where, after a fierce firefight, the battalion began withdrawing towards the south-east to Sable-sur-Sarthe. But the Allies had already bypassed Laval and the Jagdpanzers were forced to fight their way east, with just five vehicles reaching the divisional headquarters. By Friday, 15 September the number of Jagdpanzer IV tank destroyers on hand had been reduced to three and the battalion may have received a number of assault guns at that time. In November 1944 the division was allocated seventeen Sturmgeschütz IV and forty Sturmgeschütz III assault guns with most, if not all, the former going to SS-Panzer-Abteilung 17 which had used the type in Normandy.

The division took part in Operation Nordwind and on 31 January 1945 the Panzerjäger battalion reported that the two surviving Jagdpanzer IV were with the battalion headquarters while the first and second companies were equipped with twenty-nine assault guns, probably Sturmgeschütz III. Interestingly, 3.Kompanie was still able to field a single Marder III self-propelled anti-tank gun, presumably a survivor of the Normandy battles.

The division was involved in the defensive battles on the Westwall until 22 March 1945 when all the armoured vehicles were abandoned by their crews who attempted to retreat towards the town of Wiesloch on the east bank of the Rhine. In the event all but 500 men were encircled and captured in the Pfälzerwald south of Kaiserslautern.

In early April 1945 an attempt was made to rebuild the division but the personnel strength never exceeded 8,000 and it is unclear what armoured vehicles, if any, were allocated. After taking part in the defence of Nuremburg the survivors surrendered to US Army troops on May 7 1945.

Based on the chassis of the Pzkpfw IV ausf F, the Jagdpanzer IV L/48 was fitted with a sloped superstructure made up of a front plate of 60mm thickness and sides of 40mm thick armour plate. A new angled hull front, also of 60mm thickness, provided better armoured protection and until February 1944 this featured a metal bracket which held fourteen spare track links. The rounded front corners of the Jagdpanzer IV A-0 series prototype were not incorporated into the final production design. It should also be mentioned that the superstructures of these trial vehicles, of which two were completed, were made from mild steel and neither were used in combat, contrary to what has been stated elsewhere.

As the track guards were no longer easily accessible, the tools, including an axe, a long-handled shovel, a pair of bolt cutters, a track tensioning tool, a crank starter, a large spanner and a towing hook, were attached to the rear deck. On the January production vehicles, and perhaps later, the fire extinguisher was attached to the front right-hand side mudguard behind the first bracket for the hull Schürzen. On the production version of the Panzer IV/70(V) it was moved to the rear deck and this may have been a feature of late production Jagdpanzer IV vehicles. The radio antenna insulator and the gun cleaning rods were moved to the rear of the fighting compartment. A Bosch headlight was mounted on the front left-hand side fender. On the initial production vehicles two spare road wheels were held in brackets on the hull rear plate. The turret traverse mechanism was no longer required and this meant that extra space was available internally and that the traverse muffler, which was fixed to the rear plate of the Pzkpfw IV ausf F next to the cylindrical exhaust, was not needed.

The original design called for the installation of the 7.5cm L/70 gun but as these were needed to maintain production of the Panther tank, Hitler agreed that the shorter L/48 weapon could be used. The gun was mounted in the superstructure front plate and protected by a cast mantlet referred to as a Topfblende, literally a pot or jar cover. The term Saukopfblende, meaning pig's head cover, is a post-war invention.

Each vehicle was equipped with an MG 42 machine gun and this could be fired through apertures on either side of the main gun. Both openings were protected by conical covers which pivoted outwards. A number of production vehicles were completed as Befehlspanzer command tanks and these were fitted with an FuG8 radio in addition to the FuG5, the standard wireless within platoons and companies. These command vehicles are easily identified by the antenna insulator, protected by an armoured ring, fitted to the rear deck behind the commanders hatch. The Funkgerät (FuG) 8 was used to communicate with regiment or division headquarters and was usually operated with a Sternantenna D antenna, so named for its distinctive star-shape design. The first production vehicles were delivered from Vomag as Sturmgeschütz neue Art, or new model assault gun, and the name Jagdpanzer IV ausf F is not encountered in official documents until September 1944. Many of the design features mentioned here and on the following pages are also highlighted in the colour illstrations and photographs that accompany the main text. As with most weapons systems the Jagdpanzer IV was subjected to a number of changes and modifications and these are set out below.

January 1944. The first Jagdpanzer IV, Fgst. Nr. 320001, left the Vogtländische Maschinenfabrik AG (Vomag) assembly line at the company's Plauen factory (1). All Jagdpanzer IV vehicles were supposed to be fitted with the Nahverteidigungswaffe, a close-defence weapon mounted in the superstructure roof, but as adequate supplies could not be procured many vehicles left the factory with the aperture covered by a circular metal cap bolted in place next to the commander's hatch. By the end of the month a total of thirty vehicles had been accepted by the Heereswaffenamt.

February 1944. The spare tracks, which had been held on the hull front by a metal bracket, were moved to the hull rear plate. This went some way towards alleviating the problems with excess weight at the front of the vehicle and the two spare road wheels, which had been fixed to the hull rear, were moved to a position on the deck over the left-hand side ventilation screen next to the jack. Vomag was able to deliver forty-five completed vehicles during February, meaning that seventy-five Jagdpanzer IV were now on hand with the Heereszeugamt, but none were shipped to operational formations or replacement units.

March 1944. The aperture for the machine gun on the left-hand side of the hull front was covered by a circular piece of 60mm thick armour plate welded in place. During this month, and into early April, a number of vehicles were fitted with a Rundumsfeuer remotely-controlled MG 42 mounted in front of the loader's hatch (2). A total of seventy-five Jagdpanzer IV were accepted and forty-six were allocated to units with thirty-one going to the Panzer-Lehr-Division.

April 1944. The inner gun mantlet base was rounded in an attempt to save weight. Units in the field were advised to remove the muzzle brakes from the main gun. A 7.5cm L/70 gun was mounted in Jagdpanzer IV Fgst. Nr. 320162 as the first Panzer IV/70(A). By the end of the month the Heereswaffenamt had accepted 106 completed Jagdpanzer IV and ninety-two vehicles were shipped to units with the majority going to formations in the West.

May 1944. In this month superstructures without the machine-gun aperture on the left-hand side of the hull front began to appear in production. From Fgst. Nr. 320301 the frontal armour was increased from 60mm to 80mm (3). From a planned production figure of 100, a total of ninety vehicles were accepted by the Heereswaffenamt. In May 1944, thirty-one Jagdpanzer IV were allocated to operational units (4).

June 1944. To simplify construction, the rectangular box covering the filler caps on the hull rear deck was built with perpendicular sides in place of the previous angled version. It is likely that at this time Schürzen were manufactured with simple bent handles that held the plates to the hull brackets and many vehicles may have left the assembly plant without Schürzen. During this month 120 vehicles were

Notes

1. Each vehicle was identified by a unique Fahrgestelle, or chassis, number. This is usually abbreviated to Fgst. Nr.
2. No reliable figures are available regarding this modification and the guns are rarely seen in photographs.
3. This further increased the problems with the vehicle's weight.
4. All earmarked for 15.Panzergrenadier-Division which was preparing to move from Italy to the south of France.

A Jagdpanzer IV L/48 disabled in the Normandy battles. Items to note are the conical-shaped cover for the machine-gun aperture (A), the screw thread on the gun barrel for the muzzle brake (B), the angled hull front (C) with its towing eyes (D), the hooks which held the hull Schürzen(E) and the rounded air-intakes of the brake access hatches (F). This Jagdpanzer is fitted with the earlier style track links, as most of these vehicles were, with the open centre track tooth. The images below show the muzzle brake of the 7.5cm L/48 gun (G), the metal plug used to cover the left side machine-gun aperture (H) the rack on the rear hull to hold spare track links (I) and the cylindrical exhaust muffler (J).

accepted by the Heereswaffenamt, the complete planned production figure, and a total of 133 Jagdpanzer IV were shipped from the Heereszeugamt. Thirteen were sent to replacement units, forty-two going to units on the Western Front and the remainder to the East.

July 1944. Hitler ordered that production of the Pzkpfw IV tank was to be phased out by February 1945 and replaced by the Panzer IV/70(V) and Panzer IV/70(A) which were just about to enter production. In July the Heereswaffenamt accepted 125 vehicles from Vomag, the highest monthly production rate the company was able to achieve, and a total of 167 Jagdpanzer IV were despatched to units with forty-two going to formations operating in the West.

August 1944. Vomag began phasing out production of the Jagdpanzer IV in favour of the Panzer IV/70(V). Nevertheless, ninety-two examples of the former were completed and all were accepted by the Heereswaffenamt during the month. A total of 122 Jagdpanzer IV were shipped by the Heereszeugamt, mostly to units in the West to compensate for the heavy losses suffered in Normandy.

September 1944. All assembly and manufacturing firms were instructed to discontinue the application of Zimmerit by an order dated 9 September 1944 (1). The number of return rollers on the hull side was reduced from four to three and the cylindrical exhaust was replaced by two Flammentöter mufflers.

Complaints from Jagdpanzer IV crews that rain water was able to leak into the vehicle interior led to the issue of a protective canvas cover. To hold this cover in place a number of metal eyelets – five on each side, five at the front and six at the rear – were welded to the fighting compartment. Although the planned production figure for this month was sixty, just nineteen vehicles were accepted.

October 1944. Forty-six Jagdpanzer IV were accepted by the Heereswaffenamt and fifty-two vehicles were allocated to units on the Eastern Front.

November 1944. Production of the Jagdpanzer IV ends with just two vehicles accepted by the Heereswaffenamt in November. A total of 784 vehicles were assembled.

Notes

1. The great majority of the Jagdpanzer IV L/48 tank destroyers assembled would have received a coat of the anti-magnetic mine paste as just sixty-seven were built between the beginning of September and the end of November 1944 when production ceased.
2. It should be remembered that modifications were incorporated into production when the components or equipment became available and often after existing stocks were exhausted and it may have been some time before theses two changes took place.

In late January 1944 Hitler revived the idea that the Jagdpanzer IV should be fitted with the 7.5cm L/70 gun which had a greater range than the L/48 weapon. In early April a 7.5cm L/70 gun was mounted in Jagdpanzer IV Fgst. Nr. 320162 as a trial and after seeing a photograph of the resulting vehicle the Führer ordered that it should go into production without delay. It seems that little or no consideration was given to the problems that might result from the increase in weight. On 18 July 1944 Hitler ordered that the new Jagdpanzer should be named Panzer IV lang (V) but this was the source of some confusion as models of the Pzkpfw IV armed with the longer barrel 7.5cm gun had been referred to for some time as Pzkpfw IV lang, or long. Because of this, the tank destroyers were usually identified as Panzer IV/70(V), for Vomag, and this title was adopted officially in November. The new tank destroyers went into production alongside the Jagdpanzer IV and aside from some minor differences the two vehicles were identical, except of course for the main gun. A number of modifications were incorporated during production, mainly to compensate for the added weight of the L/70 gun, and these are listed below.

Notes

1. Production statistics would suggest that between sixty and seventy vehicles may have been coated with the anti-magnetic mine paste during assembly.
2. These covers, and their method of attachment, were identical to those issued for the Jagdpanzer IV.
3. See Panzer IV/70(A) below.
4. The earlier deflector had been cast as a single part.

August 1944. The first production model of the Panzer IV/70(V), Fgst. Nr. 320651, left the production lines. From a planned output of sixty vehicles a total of fifty-seven were accepted by the Heereswaffenamt. Just eleven tank destroyers were shipped in August and all went to Panzer-Brigade 106.

September 1944. All assembly and manufacturing firms were ordered to discontinue the application of Zimmerit (1). The first two road wheels on each side were replaced with steel-tyred wheels. In an effort to save production time and reduce bearing consumption, the number of return rollers on each side of the hull was reduced from four to three. One of the spare road wheels carried on the rear deck was a steel-tyred model. The early cylindrical exhaust was replaced by two Flammentöter flame-dampening mufflers. It was found that rain could easily leak into the interior and a canvas cover was issued to each vehicle (2). Forty-one vehicles were accepted and a total of seventy-six shipped to combat units, fifty-four going to units serving in the West.

October 1944. Vomag planned to complete 126 vehicles in this month but only 104 were accepted by the Heereswaffenamt. Sixty-five Panzer IV/70(A) were shipped to units, all but ten going to formations in the West.

November 1944. Three sockets, or Pilze, were fitted to the superstructure roof to hold the 2-ton jib boom which was used to remove the engine in the field. Three blunt metal spikes, one in front of the commander's hatch and two at the rear, were welded to the turret roof to hold an Entfernungs-Messer (EM) rangefinder (3). A new vertical towing coupling was introduced into production and welded onto the rear hull. Units in the field were authorised to add this modification to their vehicles. At about this time a simplified travel lock for the main gun was introduced. A total of 178 vehicles were accepted by the Heereswaffenamt.

December 1944. A new protective deflector for the Selbstfahrlafetten-Zielfernrohr sight, made up of segmented parts, was installed in the superstructure roof (4). The planned monthly production figure of 180

The first Panzer IV 70(V), which was actually a Jagdpanzer IV armed with the 7.5cm L/70 gun. Note that hull of this vehicle was built with two machine-gun apertures. Note also the earlier style gun support or travel lock with its distinctive cut-out.

This late production Panzer IV 70(V), captured intact by the Allies in North-west Europe, is painted in one of the factory-applied camouflage schemes that came into effect in late August 1944. This vehicle has three return rollers but appears to lack the Pilze sockets fitted to the superstructure roof, suggesting that it was assembled at some time in late September or early November 1944. Also note the steel-rimmed road wheels fitted to the first bogie. Shown below are the Flammenvernichter, sometimes referred to as Flammentöter, exhaust mufflers (A), the same mufflers with the Absatzkrümmer curved attachment (B), the simple handle that replaced the air-intakes on the brake access panels (C), the early type gun support (D), the later style support without cut-out (E), Scherenfernrohr 14 Z commander's periscope (F), the metal spikes which supported the EM 9 rangefinder (G) and the gunner's Selbstfahrlafetten-Zielfernrohr 1a sight (H).

was achieved and a total of 212 vehicles were shipped to operational units by the Heereszeugamt.

January 1945. Despite heavy Allied bombing, Vomag was able to complete 185 tank destroyers, the highest monthly output for this vehicle. The Heereszeugamt allocated 134 Panzer IV/70(A) in this month with almost all going to units serving on the Eastern Front or preparing to move there.

February 1945. During this month and in March the air intakes and their covers on the brake access hatches on the hull front were deleted. The cover was replaced by a simple handle welded to the hatch. The cast idler wheel, which had been incorporated into the Pzkpfw IV programme in October 1943, was introduced into production. Of a planned 250 vehicles, just 135 were accepted by the Heereswaffenamt. A total of 227, the largest monthly allocation, were shipped to formations.

March 1945. Due to continued Allied bombing the planned monthly production figure of 200 was revised to 100 but by the end of the month just fifty had been accepted.

April 1945. Ten vehicles were supposedly assembled from available parts at the Vomag plant, which was largely in ruins, by men of II.Abteilung, Panzer-Regiment 33. If this is correct then the final production figure would have reached 940.

In late June 1944 the Heereswaffenamt reported that the models of the Pzkpfw IV then in service were inferior in both penetrative power and armour protection to the Soviet T34/85 and JS122. A partial answer to this dilemma was seen in the powerful 7.5cm L/70 gun which was fitted to the Panther but which, for a number of reasons, could not be mounted in the turret of the Pzkpfw IV. The introduction of the Panzer IV/70(V) into front-line service was expected to counter the Russian designs but the production targets were rarely met and an interim solution was put forward by the firm of Altmärkische Kettenwerk GmbH (Alkett). The designers at Alkett submitted a proposal for an assault gun which would incorporate the L/70 gun and the chassis of the latest model of the Pzkpfw IV tank to supplement the production of the Panzer IV/70(V). The Heereswaffenamt agreed and calculated that such a vehicle would be superior to any enemy tank then in service except the JS122.

But by this time the Pzkpfw IV had declined in importance to the extent that it had been decided that production would cease completely in February 1945. The only company producing the tank after the summer of 1944 was Nibelungenwerke at St. Valentin in Austria which had been manufacturing the Pzkpfw IV ausf J, the last version, since June. As the fuel tanks of this model were situated directly below the turret, the Vomag superstructure would not normally have been compatible with the chassis as the gun elevation would have been severely restricted. The Alkett design solved this problem by raising the hull sides by 58 centimetres and this gave the vehicle its distinctive silhouette. A prototype was quickly built and demonstrated to Hitler, who was an enthusiastic supporter of the Jagdpanzer concept, and ordered into production as the Panzer IV lang (A), to differentiate it from the Vomag design. The name Panzer IV/70 (A) does not appear in any official documentation until November 1944 and it should be mentioned that the term Zwischenlosung, meaning interim solution, is a modern creation. Modifications introduced during production are listed below.

August 1944. The first production models left the factory assembly line at the Nibelungenwerke plant in Austria beginning with Fgst. Nr. 120301. Although it was planned that fifty vehicles would be completed in the first month just three were fully assembled and none were allocated to units (1).

September 1944. All assembly and manufacturing firms were ordered to discontinue the application of Zimmerit. Perhaps a handful of Panzer IV/70(A) vehicles, the three built in August and those assembled in the first week of September, may have received a coat of the anti-magnetic mine paste (2). Two flame-suppressing exhaust mufflers, referred to as Flammentöter or Flammenvernichter, replaced the large cylindrical exhaust. The first two, and sometimes four, rubber-tyred road wheels were replaced by steel-rimmed wheels to support the weight of the main gun. In the same month Schürzen made from wire mesh and referred to as Drahtgeflecht were introduced. Although significantly lighter, these were difficult and time-consuming to fabricate and adequate stocks may not have been available until October or even November. These are sometimes referred to as Thoma Schild or Thoma Schürzen (3), and seem to have been manufactured in a least two similar but distinct versions for the Panzer IV/70(A). It is possible that vehicles assembled in August and early September were expected to be fitted with the metal plate Schürzen designed for the Panzer IV/70(V), as the prototype was.

The planned production figure for September was 100 but just sixty Panzer IV/70(A) were accepted by the Heereswaffenamt. A total of twenty-two vehicles were allocated to units in this month with just five going to the Western Front.

October 1944. The hull sides were extended at the front and cut to form simple towing eyes and the bolted-on couplings were deleted. This modification may not have taken effect in Panzer IV/70(A) production until December. Although it was expected that 150 vehicles would be completed in October only forty-three were accepted and a total of fifty-one tank destroyers allocated to units in this month, all of which were serving in the East.

Notes

1. Those three were, however, accepted by the Heereswaffenamt and these vehicles were probably built with earlier features such as Zimmerit, the cylindrical exhaust muffler and metal plate hull Schürzen.
2. I have been unable to find a single photograph of a Panzer IV/70(A) with even a trace of Zimmerit.
3. Named for Oberst Wolfgang Thomale, the chief of staff to the Generalinspekteur der Panzertruppen

Above. A. The steel-rimmed road wheels fitted here to the first two bogies. Photographs show that vehicles were assembled with from one to four of these wheels on each side, probably depending on availability. Note that the wheels are obviously painted in camouflage colours as are the Drahtgeflecht wire mesh panels. Note also the Bosch headlight, early brake access hatches with air-intakes and the bolted-on towing coupling on the hull front. B. The tubular metal rail which supported the Thoma Schild wire mesh panels. These were introduced in September 1944 but the mesh screens may not have been available until November and this probably explains why many photographs show Panzer IV/70(A) vehicles fitted with the complicated system of brackets and rails that held the screens in place but without the actual panels. C. A close-up shot showing the brackets that were bolted to the track guard to hold the Drahtgeflecht panels in place and part of a damage mesh panel. Difficult to manufacture, they were easily damaged and may have provided very little extra protection.

Above: Production details of the Panzer IV/70(A) tank destroyer including the Vorsatz P mount and curved gun barrel (A), the armoured cover for the FuG8 radio antenna insulator (B), the hull front towing coupling (C), the spare wheel installation (D) and Flammenvernichter exhaust mufflers (E), the hull machine-gun aperture armoured cover in the open (F) and closed (G) position, the gun support or travel lock (H) and driver's visor (I).

November 1944. The Heereswaffenamt authorised the use of the Vorsatz P in production although none of these weapons were available until January 1945 (1). The Heereswafeenamt ordered that every third vehicle should be equipped with a SF14Z stereoscopic periscope and an Entfernungs-Messer (EM) 0.9m rangefinder (2). The planned allocation for November was 100 although just twenty-five vehicles were accepted. A total of forty-nine vehicles were issued to front-line units in this month with thirty-eight going to the East. The order for the despatch of the latter may in fact have been issued in the last days of October.

December 1944. In an effort to simplify production and reduce the consumption of bearings, the number of return rollers was reduced from four to three. A new vertical towing coupling was introduced into production and welded to the rear hull. In total seventy-five Panzer IV/70(A) tank destroyers were accepted by the Heereswaffenamt in December and twenty-four were allocated to units, all of which were serving on the Eastern Front.

January 1945. Three blunt metal spikes, one in front of the commander's hatch and two at the rear, were welded to the turret roof to hold the EM 0.9m rangefinder. A curved attachment for the Flammentöter exhaust mufflers, known as an Absatzkrümmer, was introduced in the assembly of the Pzkpfw IV and Panzer IV/70(V) and it is possible that they were fitted to the Panzer IV/70(A) also. The hole for the Kugellafette Vorsatz P was no longer cut in the loader's hatch. From a planned allocation of 100 tank destroyers, fifty were accepted and sixty-two vehicles were shipped from the Heereszeugamt with twenty-four going to assault gun units except for two which went to the Sturmgeschütz-Schule at Burg. It should be noted that ten of these tank destroyers were probably never delivered and they may have been diverted to other units (3). In January 1945 it was decided that production of the Panzer IV 70(A) would cease and priority be given to the manufacture of the Pzkpfw IV ausf J armed with the L/48 gun. The planned Panzer IV/70(E) programme was abandoned.

February 1945. Due to severe Allied bombing the planned monthly production was reduced to sixty Panzer IV/70(A) and in the event just twenty were accepted. A total of forty-three vehicles were assigned to units in February although it appears that just thirty-seven were dispatched of which six went to units in the West.

March 1945. Three sockets, or Pilze, were fitted to the superstructure roof to accommodate the 2-ton jib boom. Very few vehicles may have been fitted with this modification as just one Panzer IV/70(A) was accepted by the Heereswaffenamt in this month. Thirty-four vehicles were allocated in two major shipments of 7 and 16 March and all went to units in the East. Production ended after 278 vehicles have been completed.

Notes

1. The Vorstaz P was a curved barrel attachment for the MP44 which was expected to provide an inexpensive alternative to the Rundumsfeuer remotely-control machine gun. A special Kugellafette Vorsatz P mount was designed for armoured vehicles and a number of these were fitted to the loader's hatch of some late production vehicles.

2. These binocular rangefinders were produced from 1944 and used an advance system of three-dimensional visual optics. Mounted on a special tripod, the EM 0.9m should not be confused with the Scherenfernrohr scissors binoculars.

3. Similarly, the orders may have been cancelled and this would partly explain the disparity between the 285 vehicles recorded as allocated against the 278 actually assembled.

Dragon Models Ltd
B1-10/F., 603-609 Castle Peak Rd,
Kong Nam Industrial Building,
Tsuen Wan, N. T., Hong Kong
www.dragon-models.com

Tamiya Inc
Shizuoka City, Japan
www.tamiya.com

Trumpeter/Hobby Boss
NanLong Industrial Park, SanXiang,
ZhongShan, GuangDong, China
www.trumpeter-china.com
www.hobbyboss.com

Tristar
See Hobby Boss above

Italeri S.p.A.
via Pradazzo 6/b,
40012 Calderara di Reno, Bologna, Italy
www.italeri.com

Revell
Revell USA
www.revell.com
Revell Germany
www.revell.de

Voyager
Room 501, No.411 4th Village,
SPC Jinshan District, Shanghai 200540,
P.R.China
www.voyagermodel.com

Griffon Model
Suite 501, Bldg 01, 418 Middle Longpan Rd,
Nanjing, China
www.griffonmodel.com

Aber
ul. Jalowcowa 15, 40-750 Katowice, Poland
www.aber.net.pl

E.T. Model
www.etmodeller.com

Friulmodel
H 8142. Urhida, Nefelejcs u. 2., Hungary
www.friulmodel.hu

Modelkasten
Chiyoda-ku Kanda, Nishiki-Cho 1-7, Tokyo,
Japan
www.modelkasten.com
Very difficult to navigate but worthwhile.

Eduard Model Accessories
Mirova 170, 435 21 Obrnice,
Czech Republic
www.eduard.com

Master Club
www.masterclub.ru
It appears that this firm is closely associated with Armour35, a Russsian mail-order firm.

RB Model
Powstancow Wlkp.29B,
64-360 Zbaszyn,
Poland
www.rbmodel.com

Passion Models
Availble from MS Models Japan
4841-10 Nakane, Hitachinaka,
Ibaraki 312-0011, Japan
www.msmodelswebshop.jp

Hasegawa
3-1-2 Yagusa, Yaizu 425-8711
Shizuoka City, Japan
www.hasegawa-model.co.jp

As with the previous books in this series I have endeavoured to give the reader an understanding of the development and history of the Jagdpanzer IV series of tank destroyers while also providing the modeller with the inspiration to create a scale representation. In researching this project I consulted a number of works including Thomas Jentz and Hilary Doyles's books on the Jagdpanzer IV in the Panzer Tracts series, the more recent work of Joachim Baschin and Martin Block and *Jagdpanzer IV* by Janusz Ledwoch. The Panzerwrecks series of books also proved to be an invaluable resource. I would very much like to acknowledge the contributors to the Axis History and the Feldgrau forums, especially the late Ron Klages whose research on unit histories and vehicle allocations would fill many volumes, and Martin Block. I would also like to thank the modellers who very generously allowed me to publish the images of their work and, as always, I am indebted to Karl Berne, Valeri Polokov and J.Howard Parker for their invaluable assistance with the photographs and period insignia.

Although it is not possible to be certain, these vehicles may be from one of the assault gun battalions that were issued Panzer IV/70(A) tank destroyers from January 1945. Possible candidates are Sturmgeschütz-Brigaden 280 and 341 and, less likely, Heeres-Sturmartillerie-Brigade 667. All these units reported number of Jagdpanzers on hand as late as mid-March 1945. Note the vehicle in the background at right which appears to be a Bergepanzer 38, the recovery version of the Hetzer, fitted with a 7.5cm Pak 40 anti-tank gun.